Roadside Bedrock Geology from Iron Bridge through Elliot Lake, Ontario

Steven D.J. Baumann
©2020

Midwest Institute of Geosciences and Engineering

I would like to personally thank the following individuals who have made this book possible by either joining me in the field, selecting the stops, editing, or just by encouraging my love for rocks. I dedicate this book to all of you!

Sarah M. Hall

David H. Malone

Elisa J. Piispa

Sandra K. Dylka

Veronica George

Pam A. Carlton

Don J. Baumann

Boxed front cover photo was taken by the author on November 29, 2019. It is a small unvisited outcrop between sops 546-5 (p. 32-33) and 546-6 (p. 34-35), of this book, looking northeast.

Boxed back cover photo was taken by the author on November 29, 2019. The bridge over the Little White River on Route 546, looking south-southwest.

ROADSIDE BEDROCK GEOLOGY FROM IRON BRIDGE THROUGH ELLIOT LAKE, ONTARIO

STEVEN D.J. BAUMANN

Midwest Institute of Geosciences and Engineering

© 2020

www.mige-web.org

Table of Contents:

	Stop Number:	Page:
Introduction		3
Geologic Terminology		5
The Rock Cycle		11
Cross Section of the Plate Tectonics Cycle		12
Classification Chart of Plutonic Coarse Grained (Phaneritic) Igneous Rocks		13
Applicable Geologic Time Scale		14
Huronian Supergroup		15
Major Types of Faults		16
Large Scale Geologic Structures		17
How to Read a Geologic Map		18
Bowen's Reaction Series		19
Outcrop Stops: Location Maps		21
Regional Roadside Book Maps (Ontario Only)		22
Fieldtrip Stops		23
420 meters East-northeast of Eaket Lake Outcrop	546-1	24
Little White River Northeast Narrow Pass Outcrop	546-2	26
Little White River Rapids Pass Outcrop	546-3	28
North of Little White River Bridge Outcrop	546-4	30
590 meters South of Speckle Lake on ON-546 Outcrop	546-5	32
Bend on Little White River North Outcrop	546-6	34
Cliff 1700 meters South-southwest of Bloger's Lake Outcrop	546-7	36
455 meters Northwest of Cobre Lake on ON-639 Outcrop	639-1	38
1057 meters Southwest of Cobre Lake on ON-639 Outcrop	639-2	40
208 meters South of Boland River/510 meters North of Flack Lake Outcrop	639-3	42
150 meters West of No Name Outcrop	639-4	44

Table of Contents: **Stop Number:**   **Page:**

	Stop Number	Page
103 meters Northeast of Flack Lake Outcrop	639-5	46
North of Flack-Christman Lake Connection Outcrop	639-6	48
15 meters West of Lake Twenty Three Outcrop	639-7	50
580 meter South-southeast of Lake Twenty Three Outcrop	639-8	52
620 meters East-southeast of ON-639 and Gravelpit Lake Overpass Outcrop	639-9	54
380 meters South of Mitchell Road on the West Side of ON-108 Outcrop	108-1	56
640 meters South of Truck Terminal Road on ON-108 Outcrop	108-2	58
Foodland Back Lot Outcrop	108-3	60
690 meters South of Nordic Lake Outcrop	108-4	62
43 meters on Algom Nordic Mine Road from ON-108 Outcrop	108-5	64
4710 meters North of Trans-Canada 17 on ON-108 Outcrop	108-6	66
References		68
References (Maps)		70

Introduction

This book explores many of outcrops along On-546 from Iron Bridge, through ON-639, and ON-108 through Elliot Lake and ending at Serpent River, Algoma District, Ontario. Elliot Lake, is by far, the most populated municipality along this stretch with a population of 10,498 (Statistics of Canada: Census profile, 2016 census; 12.statcan.gc.ca/census-recensement/2016/dp-pd/prof/index.cfm?Lang=E). The route forms a rough triangle (p. 21), excluding Trans-Canada 17 as that has already been covered in Volume 5 of the series, Roadside Bedrock Geology along Trans-Canada 17: from Thessalon to Sudbury, Ontario (see back cover).

I am not a fan of the way traditional roadside geology books is how they are organized. *Example: Traditionally you start out at an initial location. Then you go X-distance until you see an old yellow barn (or other landmark). Then you might turn left and travel Y-distance to an outcrop.* For one thing, if you pass it you're in trouble and you may have to double back. This can be a huge time waste if the outcrops are dozens of kilometers apart; or if distance is given in miles and you need kilometers. It restricts you to a certain set order of visits, because they go from point A to B, from B to C, then from C to D, etc. It is a poor way to navigate a roadside geology book. Landmarks are not stagnant. Barns fall down. Dirt roads become paved. Road signs change. Rob's Agates and Beer gets sold, and so on. This method was fine for the 20th century, but is archaic in an era of smartphones and "global positioning system" (GPS). Fortunately, technology does advance. Almost everyone has a GPS receiver either in their vehicle or on their smartphone. It's not common knowledge, but if you directly type in latitude and longitude as below into any smartphone and most vehicle GPS systems, it will take you to the exact spot. I demonstrate this in the below two photos.

PHOTO A

PHOTO B

Here I type in the exact coordinates for my iPhone. It is our first stop (546-1). **PHOTO A:** You type in the latitude 46.34847 (no minus because you are in the northern hemisphere). Then the longitude -83.23865 (you need the minus because you're in the western hemisphere). **PHOTO B:** Then hit "search" or "enter" and it will drop you right at the spot. I record coordinates to within a several feet (a couple of meters).

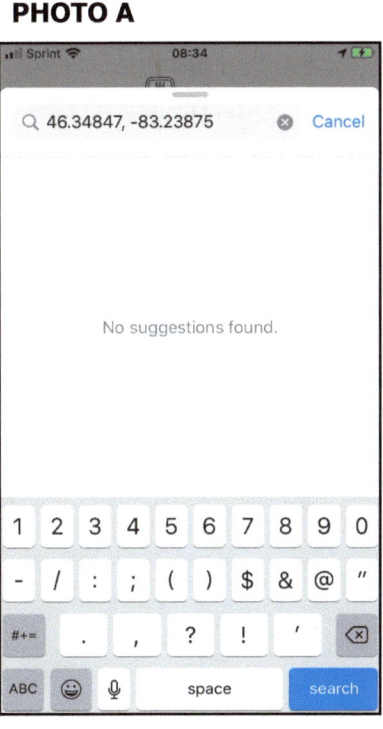
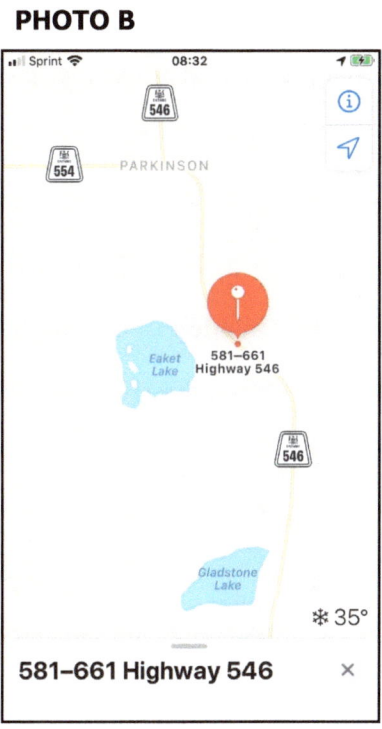

Using GPS is a big help. You can do your outcrop visits in any order that you want! You just have to remember to type the decimal latitude first then the decimal longitude with a minus in front. All the stop GPS's in this book are latitude first then longitude. Copy it into your device directly.

The book is laid out so when you open it to a stop you see all the information for all 22 stops. The basic info and geologic map on the left and the photos on the right. This way you can see all the information for a site without flipping pages.

All the outcrops in this book are on public land or in the public right-of-way. There were other outcrops that I wanted to add but didn't because you can't safely park, a better outcrop showing something similar had already been visited, or they are too clustered. I didn't want to have 10 outcrops all within 3 kilometers of one another, showing the same formation. There are also highly varying gaps between some stops, why other stops are closer together. The gaps a reflection of the local geology. The focus of this book is the bedrock. Outcrops are not consistently spaced, that's just how nature is. Also, I generally included taller outcrops so you can visit in the winter.

We visited this area November 29-30, 2019. All the photos are from those dates, and by Steven D.J. Baumann. A couple of stops are within Mississagi Provincial Park (stops 639-5 on p. 46-47and stop 639-6 on p. 48-49), but are all along the highway, so they are accessible (depending on snow cover) all year. See: ontarioparks.com/park/mississagi. The only photo not taken by Steven Baumann is on the bottom of p. 48 and is credited to Hofmann, 1971.

This book only deals with the Precambrian rocks. The reason being is that only the Precambrian forms the local bedrock. There are no Quaternary glacial stops. The oldest rocks we will visit are at least Neoarchean in age, 2,800 to 2,500 million years old (Ma). It is possible that some are older. Most of the rocks that are visited are in the first half of the Paleoproterozoic (2,500 to 2,050Ma), represented mostly by the Huronian Supergroup (p. 15). The Huronian Supergroup represents the initial rifting (igneous rocks) through the passive margin phase (mostly sedimentary rocks) of a Wilson Cycle. The Paleoproterozoic was a new time for Earth. The first free oxygen appears in the atmosphere, we have the first confirmed glacial deposits, modern plate tectonics began, and carbonate rocks begin to become more common than banded iron formations (BIF).

I hope you enjoy this book and have many delightful and exciting discoveries!

Geologic Terminology

Geologists, like most scientists, like to name things and come up with difficult to understand terms. The result can lead the average user to lose interest in the literature. I attempt to keep the technical jargon to a minimum. Sometimes it is necessary. The names of the formations, for example, are much easier to use than to describe them each and every time. When I describe the different grades of metamorphism, I avoid terms like "greenschist facies" and "amphibolite facies". Instead I use "low grade" or "high grade". It is important to have a decent understanding of the metamorphic process in the area. This section is dedicated to the explanation of terms considered unavoidable.

Banded Iron Formation (BIF)

BIF is the only sedimentary rock Earth no longer produces. All BIFs are Precambrian and are very abundant before 1.85 billion years ago. BIF is a rock where beds of an iron oxide (usually, but not restricted to, hematite) is interbedded with clastics (commonly, but not exclusively, red chert a.k.a. jasper) or carbonates.

Basalt, Gabbro, Diabase, Andesite, and Rhyolite

Basalt, gabbro, andesite, diabase, and rhyolite are all fine grained igneous rocks that are commonly associated with rifts and form directly from magma and lava. Basalt contains less than 30% quartz and feldspar and is usually dark colored. Basalt makes up most of the ocean floor. Gabbro is a coarse grained version of basalt and usually forms deep within the Earth. Andesite is similar to basalt except it tends to contain less iron minerals, like olivine. The two can be difficult to separate in the field. In the field the term diabase is used as a generically to describe a dark colored, fine grained igneous rock of variable or indeterminate composition. Rhyolite is chemically the same as granite, except it is fine grained. Most of these rocks are deposited on the surface, unless they are in dikes and sills. Since they usually form on the surface they are referred to as extrusive igneous rocks. Large intrusions that form underground tend to be coarse grained because they take longer to cool forming larger crystals, thus forming granitic rocks.

Bed

Beds are the basic divisions of layered rocks. Beds are defined by well marked surfaces called bedding planes and are where strike and dip is obtained to assess the orientation of a formation in the field. Beds can be divided into thin, medium, or thick. Beds >39.4" (100cm) are referred to as "massive" and beds <0.39" (1cm) are called laminations. Well defined continuous stacks of distinct or repeated beds are called "bed sets".

Breccia and Clasts

Breccia is used when a rock body contains angular blocks within a matrix or groundmass of different composition. Although usually reserved for igneous rocks, the term can be applied to any angular clastic rock, like sedimentary alluvial fans or debris flows. Rounded blocks are usually referred to as clasts and form the basis of conglomerates in sedimentary rocks. Clasts are also any detritus particle of any size. This can include breccia. The term "clastic rock" is used generically to refer to any rock composed of clasts. The term "clastic" excludes all rocks of biological or chemical origin (such as evaporites, carbonates, and coal).

Chert

Chert is usually a light colored rock made entirely of microscopic quartz. It is commonly formed from biological activity in the ocean. Most Precambrian chert formed differently, in that it is made of microscopic clastic quartz particles derived from wind blown deposits such as silt.

Country or Host Rock

The rock surrounding the body of an igneous intrusion. It is the rock that has been intruded, it "hosts" the intrusion. Country rock is usually used when describing the emplacement of the intrusion, in a general or regional application. Host rock is usually used when describing something in the field at the outcrop scale.

Detrital

Detrital (or detritus) or fragments of rock of any size made of lithic (rock) fragments, usually broken down by mechanical weathering.

Diamicton/diamictite

Diamicton is a relatively new term. It is essentially equivalent to glacial till. However, diamicton does not infer an origin. It is usually used to describe deposits left directly by a glacier, yet it can describe any matrix supported conglomerate. The rock form of diamicton is still referred to as tillite, although diamictite is gaining acceptance. Diamictite is the indurated term for diamicton.

Faults

Faults are a break in the Earth that forms a planar surface in which movement has occurred. The vast majority of faults have been inactive for millions (or billions) of years. Most earthquakes occur along active faults. The movement can be up, down, sideways, or a combination of any. Most faults are caused by tectonics. As rock becomes compressed or extends it can either fold (ductile deformation) or break (brittle deformation). If it breaks, a fault is created, energy is released, and an earthquake occurs. A graphic representation of the main types of faults is shown on page 17. Some faults have been so compressed that the plane on which the movement occurred is no longer easily visible. These types of faults are called shear zones.

Ga

Ga is an abbreviation for giga-annum which is used for "billions of years ago".

Granite, Tonalite, Syenite, and Diorite

Granite, tonalite, syenite, and diorite are all coarse grained igneous rocks that are commonly associated with deep magma chambers. Along subduction zones and less commonly rifts. The difference between all these rocks is the relative amounts of quartz, plagioclase, and alkali-feldspar in the rocks. Typically these rocks form deep underground and will rarely exist as dikes and sills. The term "granitic rock" is a generic textural term that has nothing to do with the actual rock type. It is just used to describe a non-foliated part of a gneiss or for any rock that looks like granite at a quick glance or any rock that expresses a granite like texture.

Great Oxygenation Event (GOE)

A period in Earth's history at about 2320Ma, when free O_2, became a permanent part of our atmosphere. Granted in extremely low concentrations.

Joints and Fractures

There are many types of breaks that occur naturally in rock where no movement has occurred. As a rock is exposed to the surface through erosion, it will often "crack". Cracks can also form in the subsurface. These cracks are called fractures or joints. Although the two are different, herein they are treated as the same feature to avoid getting too technical. Rocks can also form these features from tectonic stresses. They may also form if a melting glacier retreats. As the weight of the ice is removed the rocks near the surface will often form fractures. Sometimes the fractures become filled with other minerals and close.

Lava

A volcanic surface flow, in either the solid or liquid state, that originated from an underground magma.

Ma

Ga is an abbreviation for mega-annum which is used for "millions of years ago".

Magma

An underground melt (liquid) that forms rock once it cools to a solid. Unlike the term "lava" we do not use the same term for subterranean magma that has solidified into rock. This is a hold over from before the days we knew rocks like granites were derived from melts. Lava has been seen at the surface since humans evolved.

MSL

Mean sea level

Matrix, Cement, and Groundmass

The matrix is what binds the particles of rock together (in sandstones) or the fine grained portion of any sedimentary rock containing significant coarse grains. In sandstones the matrix is often the same as the cement (what holds the grains together). The most common minerals that make up the matrix or cement in a rock are calcite and quartz, although other minerals and small grains can serve as matrix. Cement is not used for igneous and metamorphic rock. The term groundmass is often used interchangeably for matrix, but this is not entirely correct (even I have been known to swap the two). Groundmass usually pertains to fine grained igneous and metamorphic rocks and is generally only used if the rock is porphyritic or has isolated large crystals.

Metamorphism

Rocks that have been metamorphosed have undergone deep burial and have been altered through heat and pressure but not enough to melt the entire rock. The greater the heat and pressure, the more likely the rock is to become altered and form different minerals from existing ones. The vast majority of the Paleoproterozoic rocks in along the north shores of Lake Superior and Lake Huron are metamorphic but they have undergone low grade metamorphism. This means that they were buried deep enough to become metamorphosed but not enough to have altered their mineral make up. The Archean rocks were buried far deeper and longer than the Paleoproterozoic rocks. They tend to have higher grades of metamorphism. Some high grade metamorphism has occurred near the many extinct faults in the area. Metamorphism also occurs in close proximity of dikes. Metamorphism that occurs locally near faults and intrusions is often called contact metamorphism.

Parent Rock (or protolith)

Parent rocks are the sedimentary or igneous rocks that exist before undergoing metamorphism.

Phenocryst

Phenocrysts are numerous large and obvious crystals within the groundmass of a rock, and are usually a different color than the groundmass. They are significantly larger than the next discernable crystals. There really is no standard for what constitutes "significantly larger". If I can see the crystals from 8-10 feet (a couple of meters) away, then I consider the rock porphyritic (the adjective used to describe a rock with phenocrysts).

Pluton

An igneous body that cooled slowly underground, allowing for relatively large crystal growth. A term typically used for granitic rocks.

Quaternary

The Quaternary is the period of geologic time from 2.4 million years ago to the present. It encompasses all of the modern ice ages. Quaternary deposits are common in the area but rarely exposed. They do commonly leave scour marks and striations on the Precambrian rocks.

Rifts, Passive Margins, and Wilson Cycles

Where the earth's crust begins to rift apart (split) due to upwelling magma chambers from deep in the Earth. If the rift persists eventually a continent will split and a new ocean will form between them as they move further and further part. Not all rifts form new ocean crust. The Mid-continent rift (under Lake Superior) is a failed rift that formed around 1.1 billion years ago. The initial and early stage of rift formation is due to active volcanism. Rifting usually occurs on the continents, but can occur at the bottom of the Ocean, like the modern "East Pacific Rise". As the rift matures, volcanism on the continent stops and a passive margin forms. The passive margin becomes a place for thick sequences of marine sediments to accumulate until another continent or island arc merges with the passive margin. The force that splits and moves continents around the globe is called Plate Tectonics. A Wilson Cycle (named after the Canadian geophysicist John Tuzo Wilson, 1908-1993) is the name given to the opening of an ocean basin and its inevitable closing, including all the steps in between. This includes rifting, passive margin development, subduction along the passive margin, and the accretion of island arcs or continents as the ocean plate shrinks due to subduction.

Sandstone and Quartzite

Sandstone is the rock version of sand. Sand is sediment and sandstone is a sedimentary rock. It consists of grains of sand ranging from 0.2 micrometers (about as small as the human eye can see) to 2 millimeter in diameter (roughly the width of two fingerprint lines). Sandstone is derived from other rocks that have been weathered from larger rock bodies. Sand can come from any rock. Feldspar rich sands are called arkoses. Quartz rich sands are what people typically think of when we say sand. Sand that consists of fragments other than feldspar or quartz are referred to as lithic sands. Sands with a lot of fines (silt or clay) are called wake. Sand that is made up of mostly sand-sized particles are called arenites.

Quartzite is the metamorphic version of sandstone. Quartz rich quartzite is just called "quartzite". Arkosic quartzite is called "proto-quartzite". Quartzite rich in lithic fragments is called "immature quartzite". Silty or argillite rich quartzite are referred to as impure quartzite, or impure proto-quartzite, or impure immature quartzite.

Shale, Mudstone, Argillite, and Slate

Shale is a fine grained rock (smaller than sand) mostly made of clay. Mudstone is a sandy or silty shale. Argillite and slate are metamorphic versions of shale and mudstone.

Strike and Dip

This is a the basic measurement of rocks in the field and is the key measurement in the field. Strike is simply the line between two points of equal elevation always measured from north. These points can be hundreds of feet, several meters, or a couple of inches or centimeters apart. Dip is always perpendicular to strike in the down-slope direction given in degrees from the horizontal. When you combine the strike with the dip you get the true orientation or trend of a rock unit. The measurement is obtained with a Brunton compass and give in quadrant format always oriented to the north. A strike of 280° would be written as N80W. Dip, by definition has to be either NE or SW if your strike is NW because dip is always at a 90° angle down-slope to strike. If your strike is NE than dip has to be either NW or SE (see below for examples). Strike and dip can be used to record the orientation of any planar surface. This is usually bedding planes, and in this book it will be bedding unless otherwise noted. Strike and dip can also be used to measure fault planes, joint faces, fracture planes, and dike orientation.

To the left is a graphic depiction of strike and dip. The long line without a number attached to it is the strike (in this case it trends N80W). The long line is oriented to the strike but contains no numbers. The smaller line indicates the dip with the degree of dip from the horizontal indicated by a number. In this case the rock dips 35° to the NE, and the degree (°) symbol is dropped. We would write out this trend or orientation in this example as **N80W 35NE**.

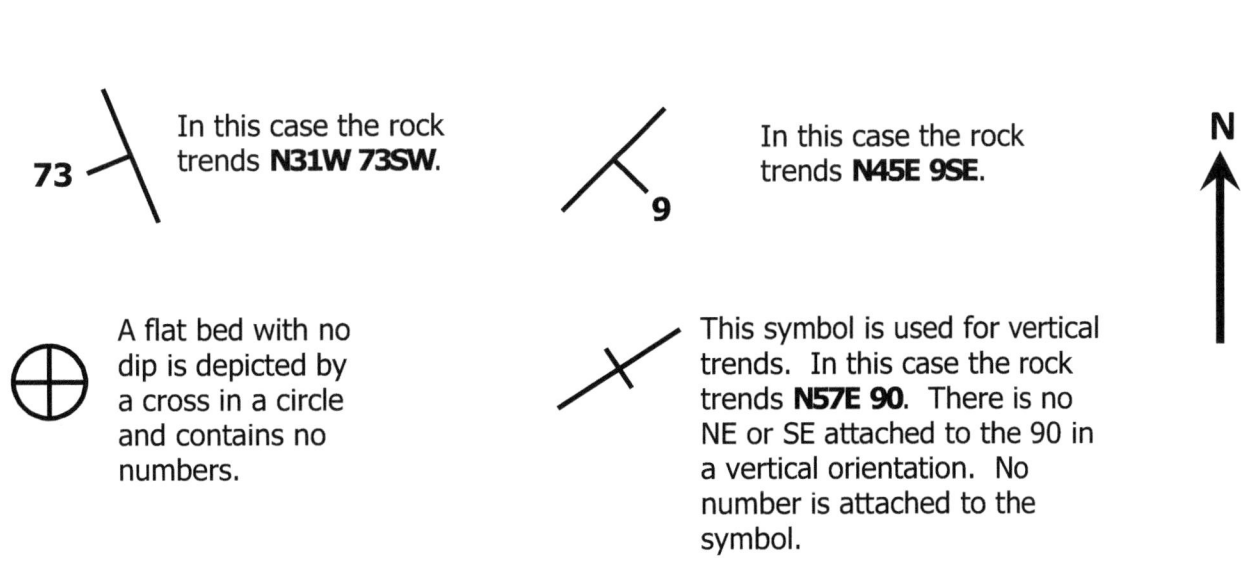

In this case the rock trends **N31W 73SW**.

In this case the rock trends **N45E 9SE**.

A flat bed with no dip is depicted by a cross in a circle and contains no numbers.

This symbol is used for vertical trends. In this case the rock trends **N57E 90**. There is no NE or SE attached to the 90 in a vertical orientation. No number is attached to the symbol.

Subduction Zones

Subduction zones form where two plates collide and one subducts under the other, retuning it to the mantle (the layer in the Earth beneath the crust and lithosphere). When two continents collide only the ocean crust is subducted into the mantle. Continents are lighter and more buoyant so they do not get subducted. As ocean crust slides under a continent on its way back into the mantle, volcanoes form. Some will form on the continent. This is currently happening along the west coast of North and South America from Alaska all the way to Argentina. Sometimes volcanoes will form on the ocean crust and create island arcs. Japan is an island arc. Eventually island arcs are pushed against a continent and become a part of it. This is essentially what occurred during the Penokean Orogeny in Wisconsin, the Upper Peninsula, Minnesota, and Ontario about 1.895 to 1.822 billion years ago.

Tonalite-trondhjemite-granodiorite (TTG)

Tonalite and granodiorite appear on the QAPF diagram, so they will not be elaborated upon here. Trondhjemite is essentially tonalite where all the plagioclase is oligoclase with little to no anorthite. TTG's usually exist as massive plutons in Hadean up through Neoarchean igneous complexes. Prior to the Mesoarchean TTG's were common globally and true granite was rare to non existent. By the beginning of the Paleoproterozoic TTG's became exceedingly rarer as Plate Tectonics took over and granite plutons became the norm. TTG's are still deposited today but are restricted to ophiolites and volcanic arc batholiths. No one knows why the Earth now favors granites over TTG's . It likely has something to do with the Plate Tectonics cycle which was not in operation during the Hadean and most of the Archean.

Vermiform structures/fossils

Vermiform structures are primary structures that have a worm-like shape to them. However, they do not have to be formed by living creatures. Tectonic processes as well as certain sedimentary environments can form certain types of vermiforms through inorganic processes.

Xenolith (or Xenocryst)

Large foreign inclusions in igneous rocks of older igneous, sedimentary, or metamorphic rock that is brought up from depth with rising magma but does not get melted. They are usually composed of the country rock but can be mantle fragments. They can be angular or rounded.

The Rock Cycle

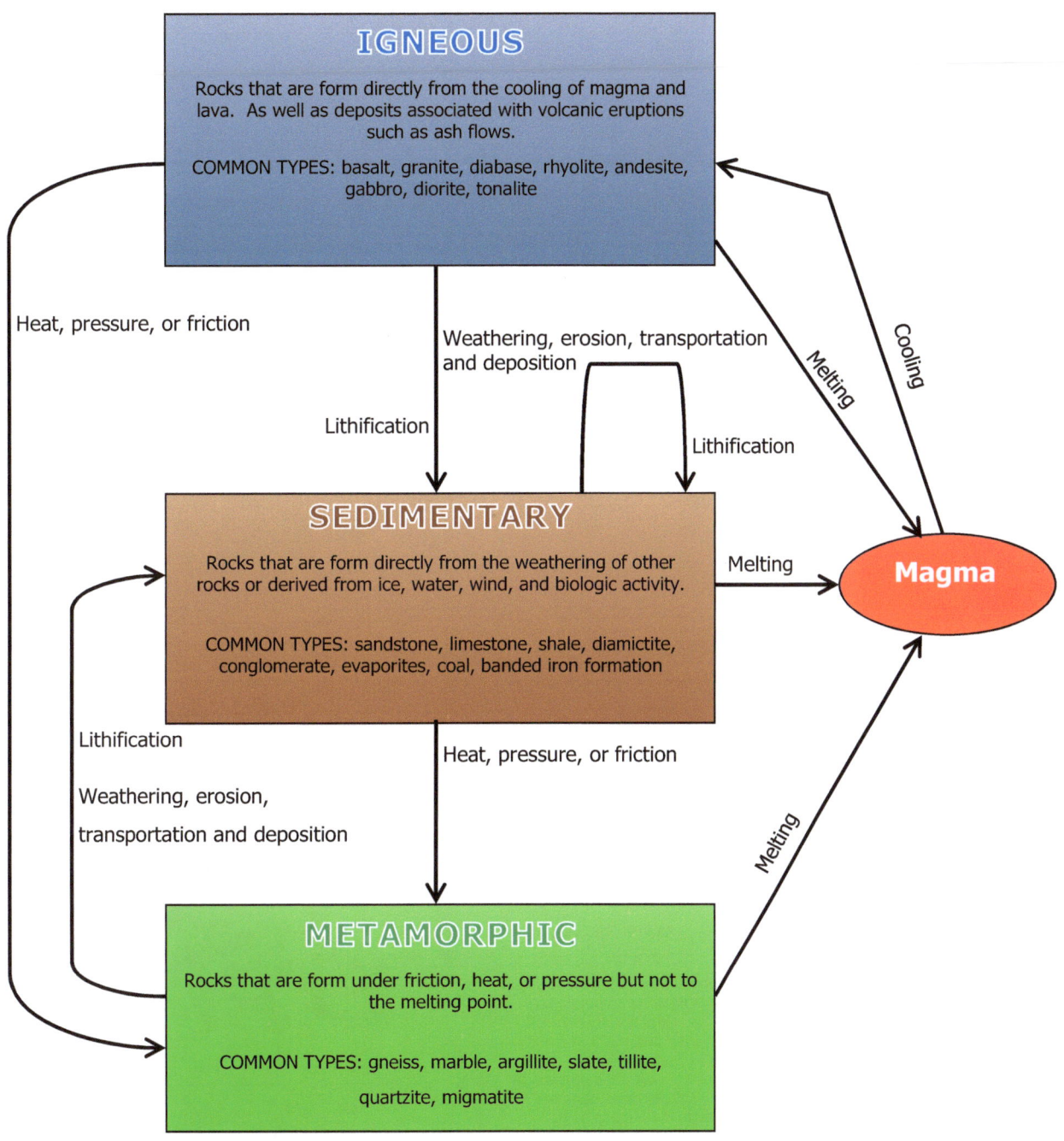

The rock cycle is the process of how one rock type becomes another. The above chart illustrates how the process unfolds. Not all rocks will become a different type, some will repeat the process again and again. The process can end anywhere on the chart. Plate Tectonics and the water cycle are the main driving mechanisms behind the rock cycle.

Adapted from Baumann (2010)

Cross Section of the Plate Tectonics Cycle

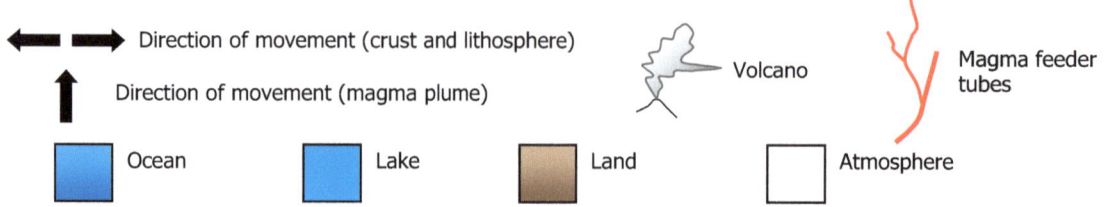

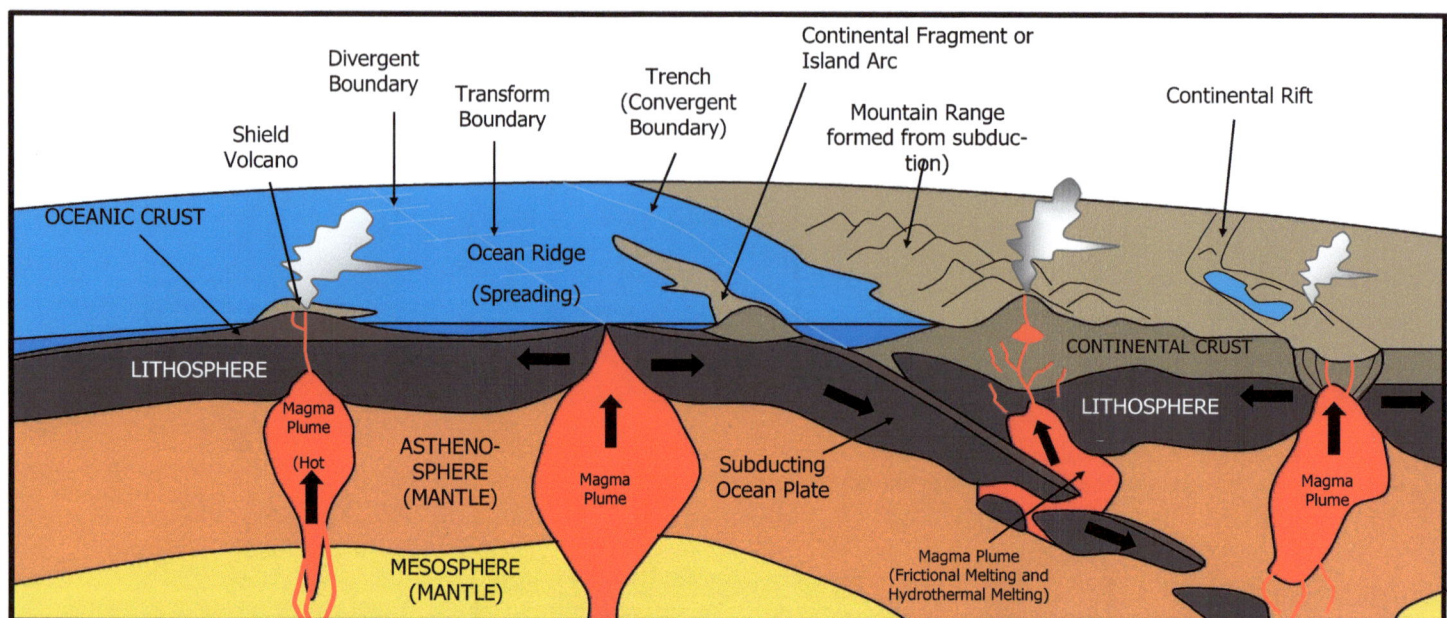

The Earth's Interior

Plate Tectonics is main mechanism of how Earth loses its internal heat. Modern Plate Tectonics has been in effect for about 2.5 billion years (Earth is 4.6 billion years old) and will likely continue for a billion more. Prior to Plate Tectonics Earth likely lost its internal heat in a similar way that Venus does with a transitional period from about 2.5 to 3.2 billion years ago.

The inner core is a solid ball of iron and nickel, and its temperature is about 10,500°F (hotter than the surface of the sun). The inner core loses heat through conduction. The exact nature of the inner core is still largely unknown.

The outer core is liquid iron and nickel and is an average of 8800°F. It loses its heat to the mantle through convection, like boiling water in a pot.

The mantle is very different. It's temperature varies greatly at about 950°F to 1600°F. It makes up most of the mass of Earth. It is not liquid as most people think. It is solid. The asthenosphere contains melted magma plumes but overall the mantle behaves like silly putty. You pull it slowly and it flows, pull it fast and it snaps. It is mostly made of the mineral olivine. Like the outer core, the mantle transfers heat to the surface through convection. However, the process is very uneven and much slower than in the outer core. This leads to magma plumes near the core-mantle boundary making their way to the surface as hot spots. It also leads to cool solid parts as subduction brings the lithosphere into the mantle. The asthenosphere is now suspected to contain hydroxide (OH^-) bounded to minerals (like ringwoodite) and derived from water.

The lithosphere and crust operate as one unit. This is where the "plate" part in Plate Tectonics comes in. As a result of uneven mantle convection, the Earth's surface is like a cracked eggshell. It is divided into 12 large plates that move around the globe as magma from the mantle rises upward (divergent boundaries) causing the lithosphere to move laterally around the globe. In other areas the lithosphere is being subducted (convergent boundary) back into the earth, where it is reincorporated into the mantle, and the process starts over.

Gravity plays a major part in the plate tectonics cycle. As ocean crust ages, it becomes more dense and begins to sink into the mantle forming a subduction zone.

Cross Section of the Earth

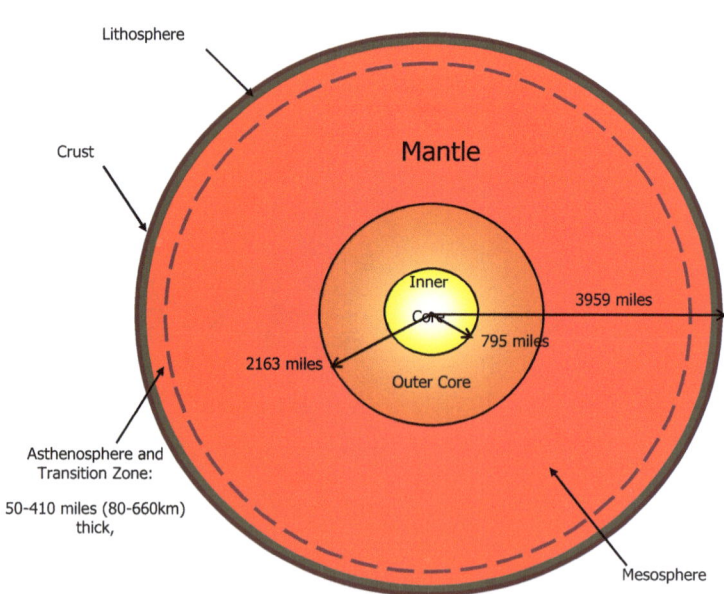

PERCENT OF MASS (EARTH = 100%) AND AVERAGE DENSITY

Unit	Mass	Density	
		(g/cm³)	(lb/ft³)
Lithosphere + Crust	= 2.2%	= 2.8	= 174.8
Mantle	= 68.3%	= 3.3 - 6.0	= 206.0 - 375.6
Outer Core	= 27.5%	= 10.4	= 649.3
Inner Core	= 1.8%	= 13.3	= 830.3
Water	≤ 0.2%	= 1.0	= 62.4

Classification Chart of Plutonic Coarse Grained (Phaneritic) Igneous Rocks

This chart is used for coarse grained (and usually plutonic) igneous rocks. Plutonic rocks tend to be coarser grained because they cool underground and thus crystalize more slowly than surface eruptions. This allows larger crystals to grow. It can be used in the field or the lab. Coarse grained in this case means the grains are easily visible to the naked eye. This corresponds roughly to 1/4 (0.25) millimeters (~0.01 inches) or the #60 U.S. and Tyler sieve sizes. This is the border between what geologists call fine and medium sized sand grains (based on the Wentworth Scale).

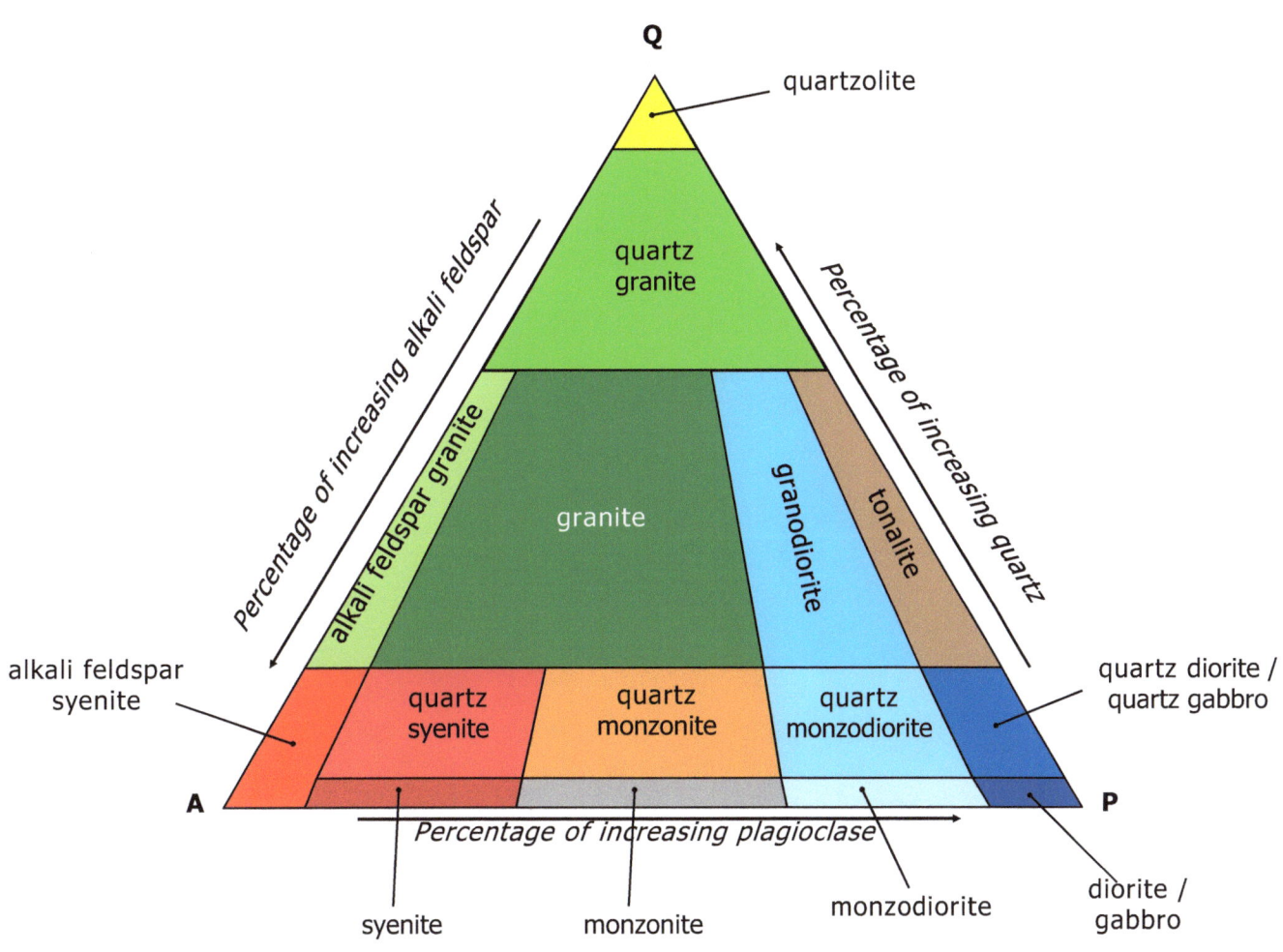

A = alkali feldspar

P = plagioclase

Q = quartz

This chart ignores the dark colored minerals in coarse grained igneous rocks, such as hornblende and mica. It deals only in the ratios of quartz, plagioclase, and alkali (a-k) feldspar.

Adapted from Streckeisen (1974)

Applicable Geologic Timescale

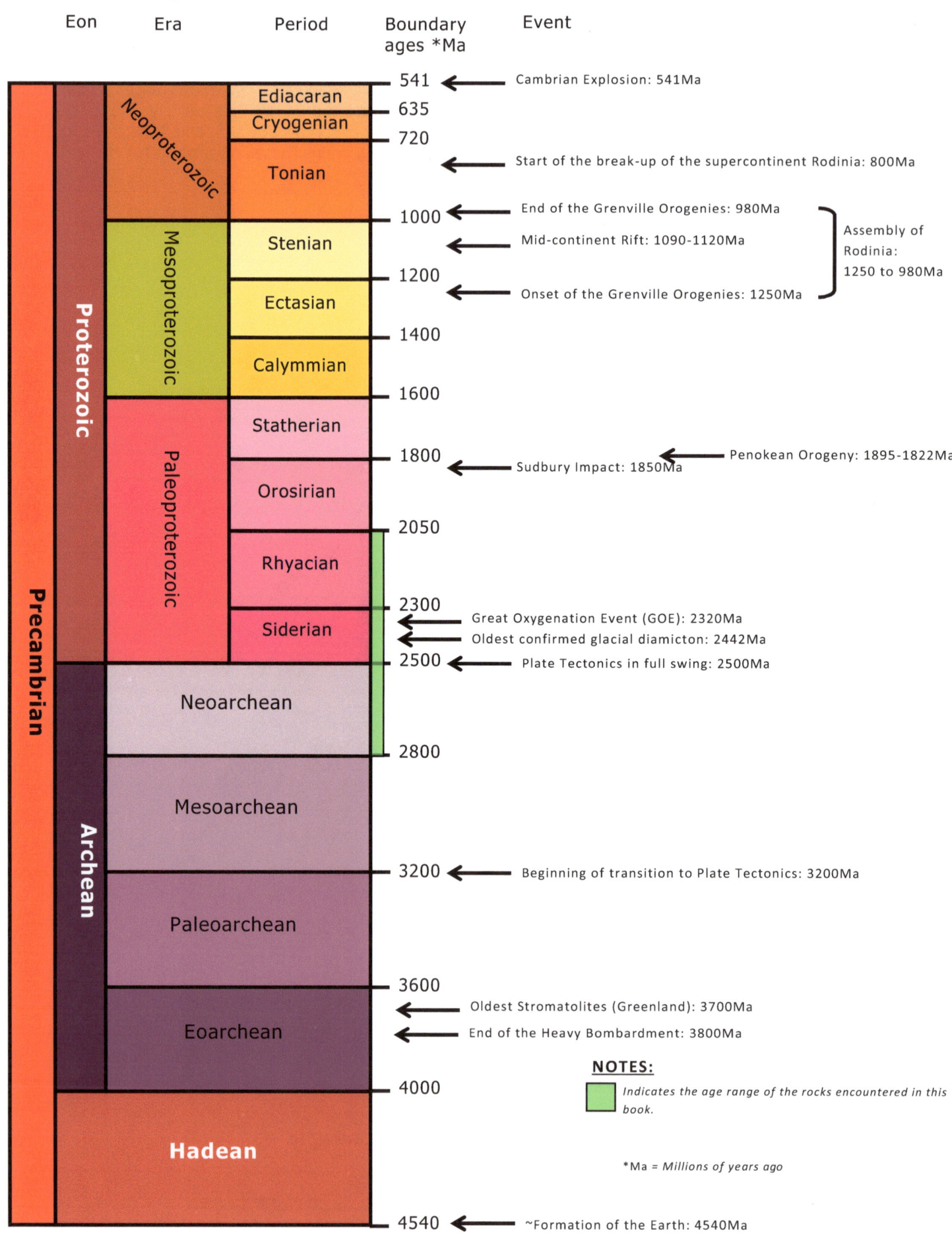

Huronian Supergroup

Almost all the rocks visited are part of the Huronian Supergroup. This Supergroup consists of 4 groups and 20 formations. The basal Elliot Lake group represents the formation of a continental rift. The Hough Lake Group represents the sediments that filled the rift zone as the ocean moved in. The Quirk Lake and Cobalt Groups represent the passive margin phase, similar to the modern North American Coastal Plain. The Nipissing represents intrusions that penetrated the Huronian as the passive margin became an area of active subduction. Not only does the Huronian demonstrate a modern style of Plate Tectonics, but it records the first free O_2 in the atmosphere as indicated by the appearance of the first "red beds" in the Firstbrook Member of the Gowganda Formation.

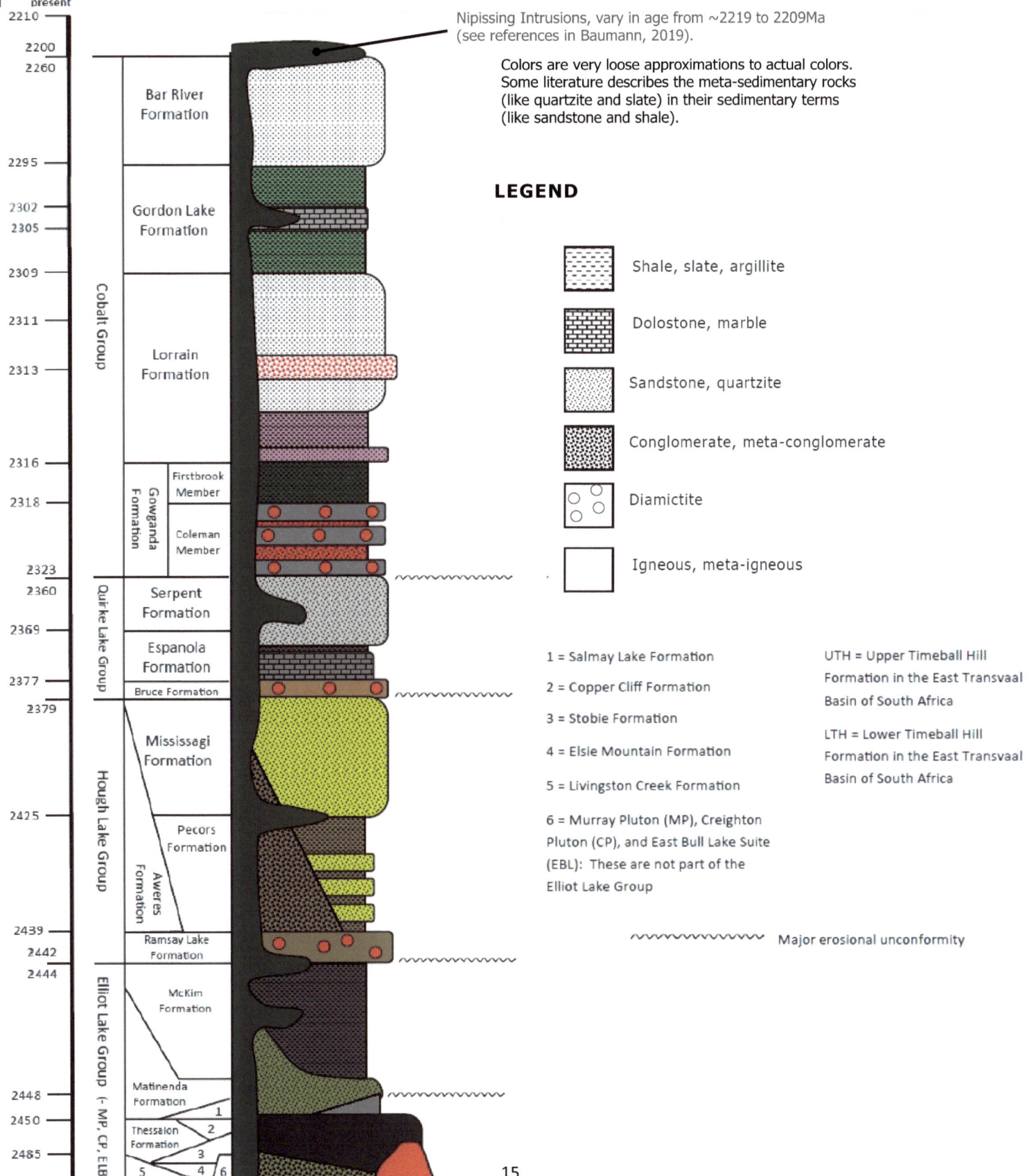

Major Types of Faults

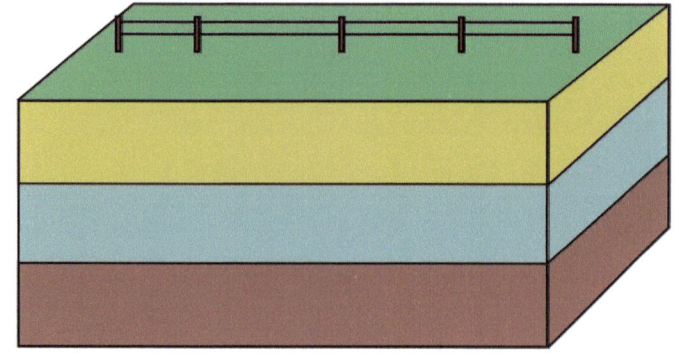

Non faulted Block

This is a block diagram of a small section of the Earth. The vertical brown posts attached with dark lines represent a fence. The green is grass. The colored layers represent strata of different buried rock types.

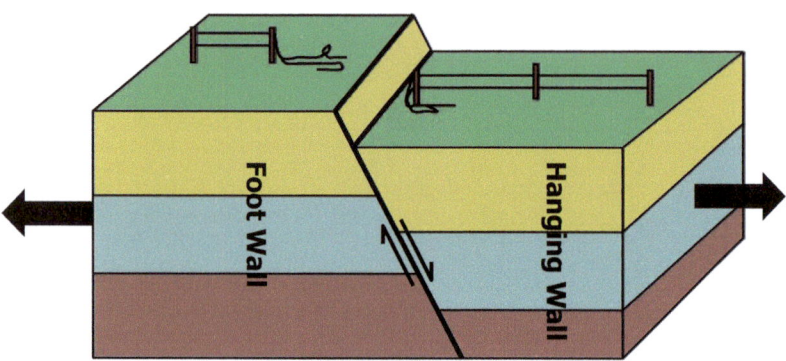

Normal faulted Block

A normal fault forms when the Earth's crust extends, causing the hanging wall to drop down relative to the foot wall. Note the position of the broken fence. The small arrows indicate relative movement. The large arrows indicate the direction of extension.

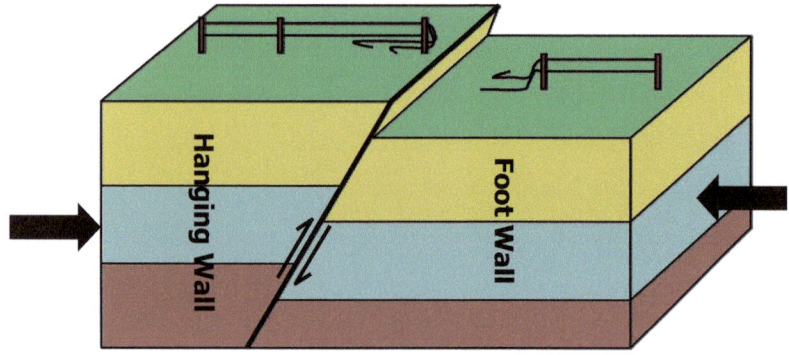

Reverse faulted Block

A reverse fault forms when the Earth's crust compresses, causing the hanging wall to move up relative to the foot wall. Note the position of the broken fence. The small arrows indicate relative movement. The large arrows indicate the direction of compression.

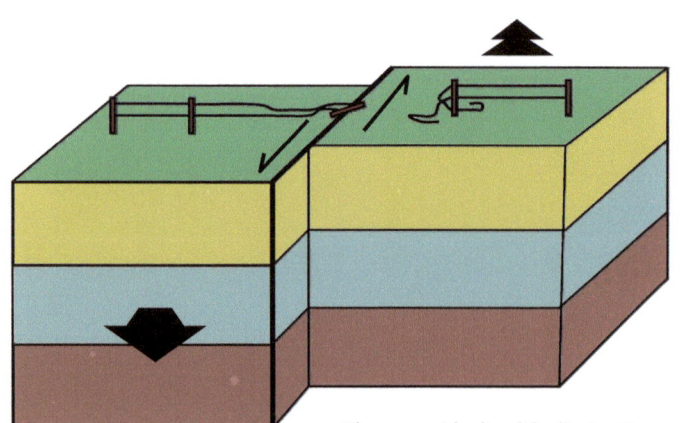

Strike-slip faulted Block

A strike-slip fault forms when one block of the Earth slides laterally passed another due to transform movement. There is no hanging or foot wall. Note the position of the broken fence. The arrows indicate relative movement.

These are idealized fault situations and are only the three main types. Most faults are far more complex

Large Scale Geologic Structures

Large scale geologic structures are mechanical alterations in the rock on a scale larger than a person. Some large scale structures, such as faults, can also be small scale. They can take up less than one square mile or thousands of square miles. They form after the sediment or rock has been deposited. They can be formed by tectonic movement, erosion, or intrusive magmas. The diagram below is a hypothetical cross section, a slice through the Earth's crust. The colors below represent different types of rock.

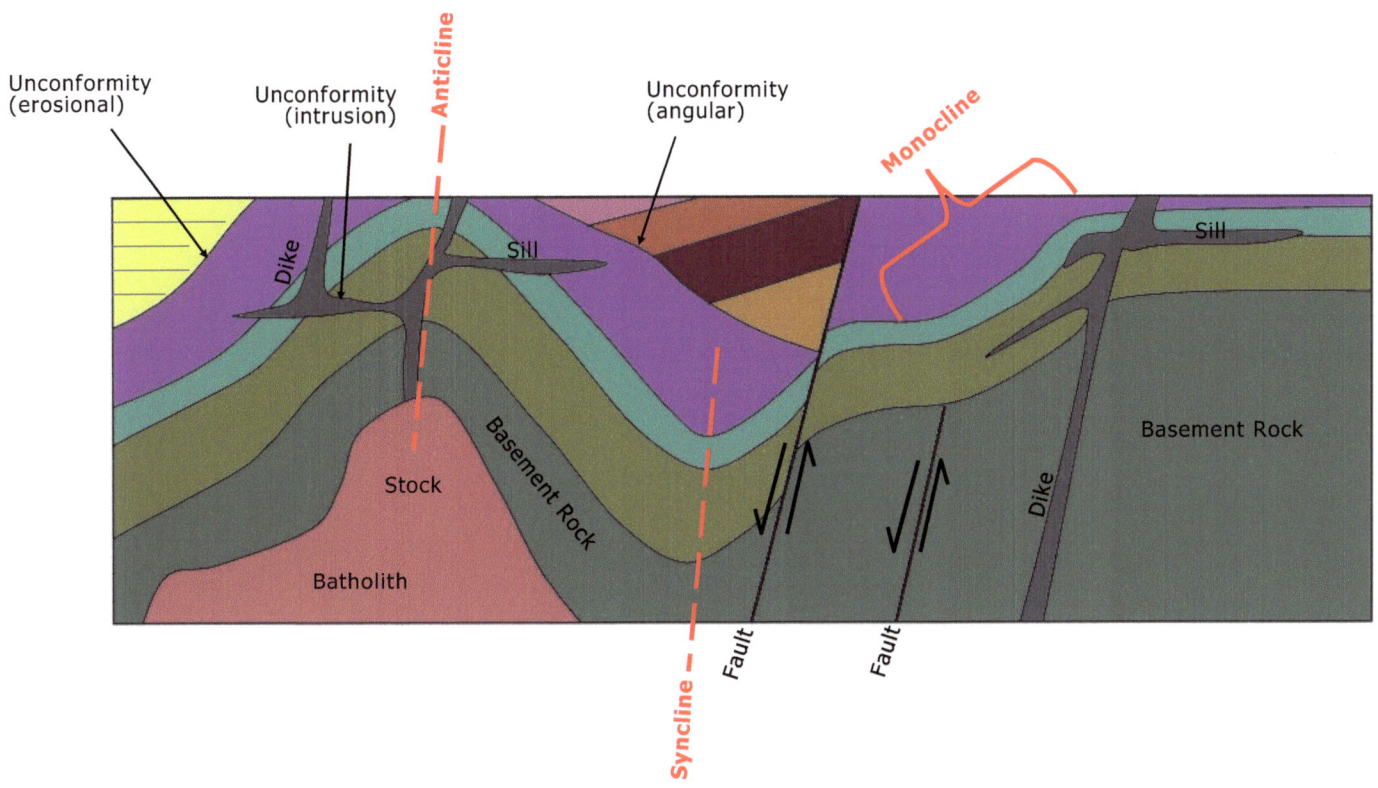

Anticline: A folded set of rocks where an open ended ridge forms. This may or may not be reflected in the surface topography. Older layers are towards the center. An anticline that forms the shape of an upside down bowl is

Basement Rock: Usually refers to the oldest (Precambrian) rocks in an area. They can be anywhere from deep underground to at the surface.

Batholith: A large igneous intrusion (usually several square miles) that was originally deposited by magma deep

Dike: A vertical, narrow, igneous intrusion that has cooled from magma within existing rock.

Fault: A fault is a planar surface, or crack, within the Earth along which movement has occurred. If a crack forms

Monocline: A folded set of rocks, resembling a step, where only one end is steeply folded in the middle and the surrounding rock is relatively flat. It may be part of an anticline or syncline.

Sill: A horizontal, narrow, igneous intrusion that has cooled from magma within existing rock.

Stock: A small igneous intrusion (usually around one square mile or less) that was originally deposited by magma deep underground. Usually a coarse grained granite type rock. Stocks are often connected to batholiths.

Syncline: A folded set of rocks where an open ended depression forms. This may or may not be reflected in the surface topography. Younger layers are towards the center. A syncline that forms the shape of a bowl is closed and referred to as a basin.

Unconformity: An unconformity represents a missing block of time in the rock record, usually due to intrusions, erosion, or non deposition. There are many different types. An angular unconformity is the easiest type to spot in the field. Angular unconformities have one set of rocks at an angle to another set of rocks.

How to Read a Geologic Map

Geologic maps are produced in order to represent geologic units on a base map (usually on a topographic map but can be any map) of a selected area. The purpose of a geologic map is to decipher and predict the relationships between geologic units in the selected map area. Most geologic maps represent the geologic units with colors (although some are black and white) and abbreviations of geologic units, which are explained in the legend. The map below is an excerpt from a map created by the author in 2013. The legend explaining the geologic units is not shown to conserve space. Maps presented in this book are accompanied with legends. Legends include not only the map symbols, but also geologic units with names and descriptions. Features typically found on geologic maps of the area are shown below with arrows and circles. All geologic maps must contain three things. 1-north arrow. 2-scale. 3-map legend. The United States Geological Survey has set standards for geologic symbols such as contacts, strike and dip, faults, anticlines, and other structures not depicted below. Colors are also standardized. However, if you have a lot of geologic units, especially when most of them are within the same span of geologic time, the colors would be so close to one another that it is acceptable to deviate from the standard. For example, if you had 13 geologic units, all of which are Paleoproterozoic, it wouldn't make sense to have 13 shades of purplish red on your map.

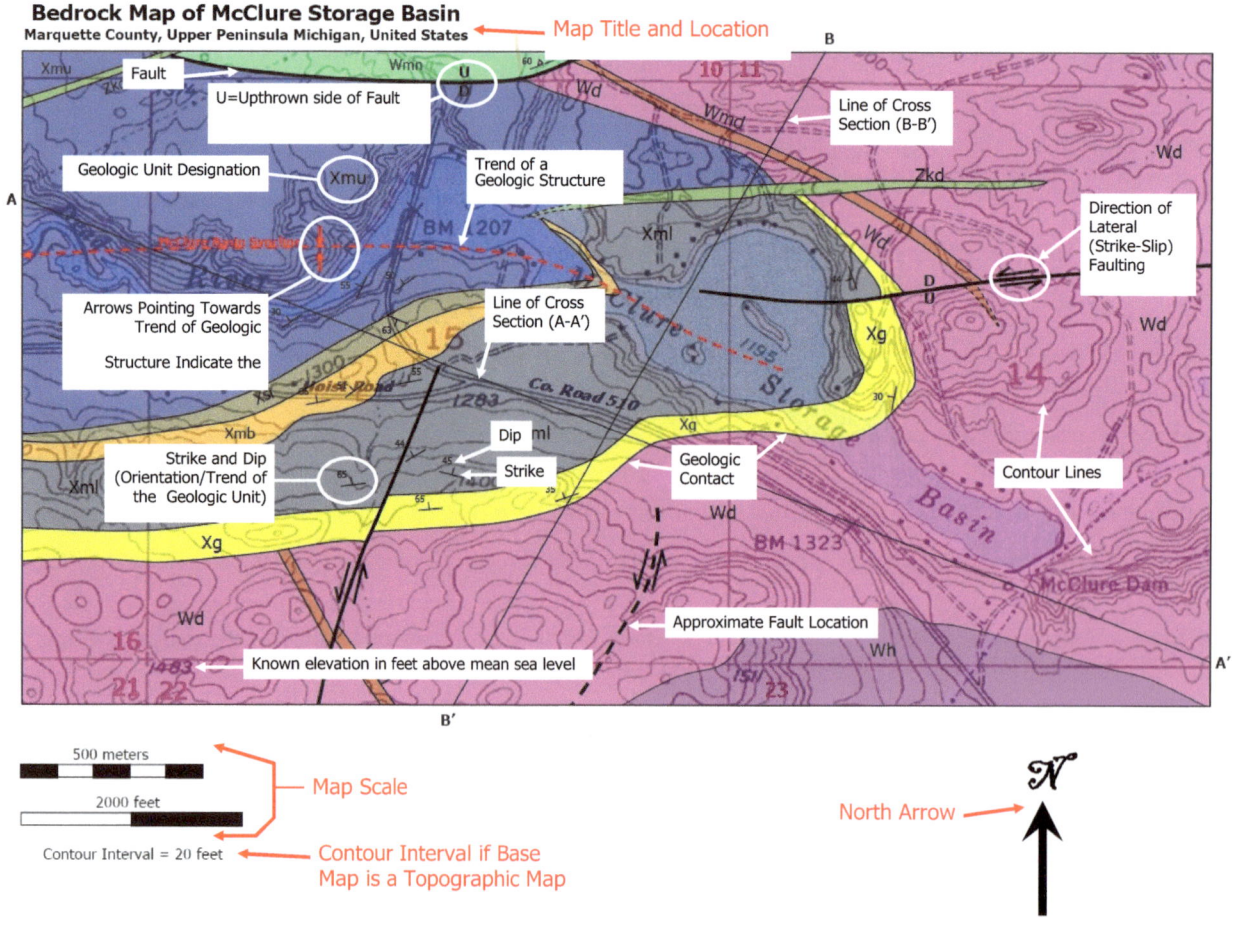

Bowen's Reaction Series

This is an important chart that illustrates how minerals form as they cool from magma. Magma is not made of minerals. It is a chemical liquid from which minerals form. Minerals by definition are solids. Bowen's reaction series chart assumes a magma body with all the available elements to form minerals on the chart as the magma cools. If an element weren't present, say, iron, you would not get olivine. The chart is an idealized reaction series. It does not take into account things like two chemically different magmas merging, rapid verses slow cooling, fragmentation of the magma, chemical enrichment of the magma due to hydrothermal or gas injection, or incorporation of abundant and chemically distinct country rock into the magma as it rises and cools.

At the dawn of the 20th century, Norman L. Bowen and others, began experimenting with rocks in order to see if certain minerals would crystalize from a magma first. Several things were discovered out of the experiments. First, the magma (or melt) would try to stay in equilibrium with the forming crystals (or minerals). This would result in different compositions between the melt and minerals, and would through off equilibrium. So the new crystals would re-react with the melt to form new minerals. Second, the newly crystalized minerals would form in a specific order. Third, if the magma had enough silica and was homogenous, two main series would form. The continuous and discontinuous series, which would both merge to a simpler chemical melt composition of the magma as the iron, magnesium, calcium, sodium, etc. was used up to form the residual series.

The continuous series deals with the crystallization of the feldspars. First plagioclase minerals will form. As the magma cools, this will through off the equilibrium. So some of the calcium in the plagioclase will re-react with the magma and become potassium/sodium enriched until finally orthoclase crystalizes out. The discontinuous series is odd...but makes sense. Say the magma produces olivine at a high temperature. As it cools further the olivine reacts with the melt but doesn't exchange ions like in the continuous series. Instead it will "change" into pyroxene. The chemical reaction would look something like this (from a generic olivine to pyroxene): $Mg_2SiO_4 + SiO_2 \longrightarrow 2MgSiO_3$. So, instead of say calcium being replaced with potassium; the added silica causes the existing mineral to reorganize itself but made of the same chemicals. Basically all that is occurring is there is an internal crystal lattice (or structural) adjustment to achieve crystalline stability at lower temperatures until orthoclase/microcline is formed. Whether continuous or discontinuous, once orthoclase/microcline forms, only the residuals remain. Muscovite will form until everything but silica (SiO_2) is left to form quartz.

How does this help us identify minerals? For starters, olivine and anorthite can form together as they are both mafic type minerals. But pyroxene and quartz cannot form simultaneously since one is mafic type and the other is felsic type. Say you have a fine to medium grained igneous rock and you see about 50% very dark but unidentifiable mineral, and 50% nearly white and heavily striated mineral. You can't identify the dark mineral but it can be assumed to be pyroxene. Why? Because the other mineral is white and heavily striated, so it has to be anorthite. Be careful. Just because anorthite and quartz can't form at the same time, that doesn't mean they can't be in the same rock as quartz can still form at low magma temperatures if the magma is saturated in silica.

As ingenious as Bowen's reaction series is, it cannot account for everything. In Bowen's day, we didn't know about plate tectonics. The rolls of saturated verses unsaturated magma, incorporation of country rock, differentiation of magma due to cumulate was only beginning to be studied. During his time people were still arguing if granite was igneous or metamorphic. Bowen's reaction series is very useful but it deals with idealized conditions. This is why field mapping in conjunction with laboratory petrographic analysis is so important. The two cannot exist in isolation.

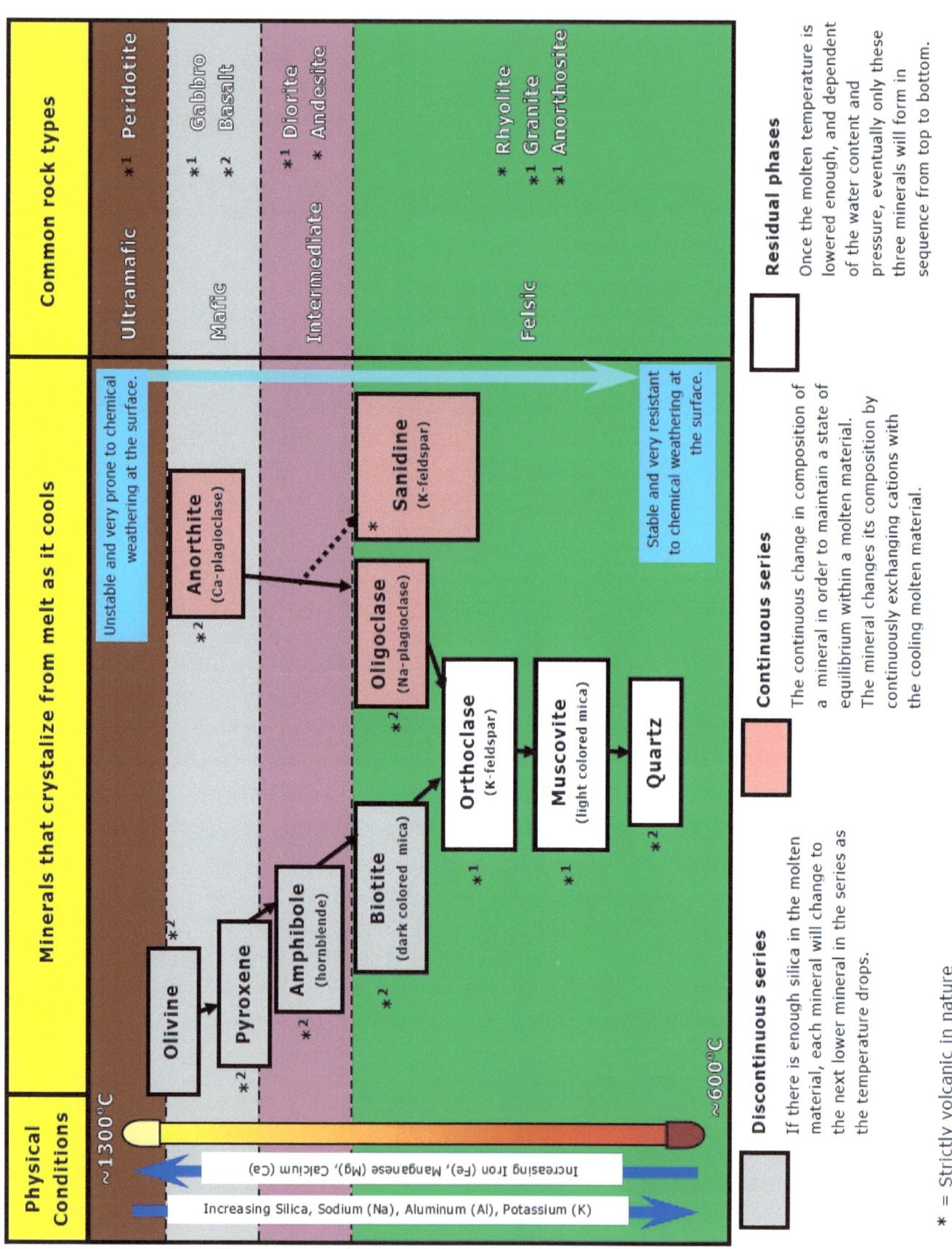

*Adapted from Bowen (1922) and others

Outcrop Stops: Location Maps

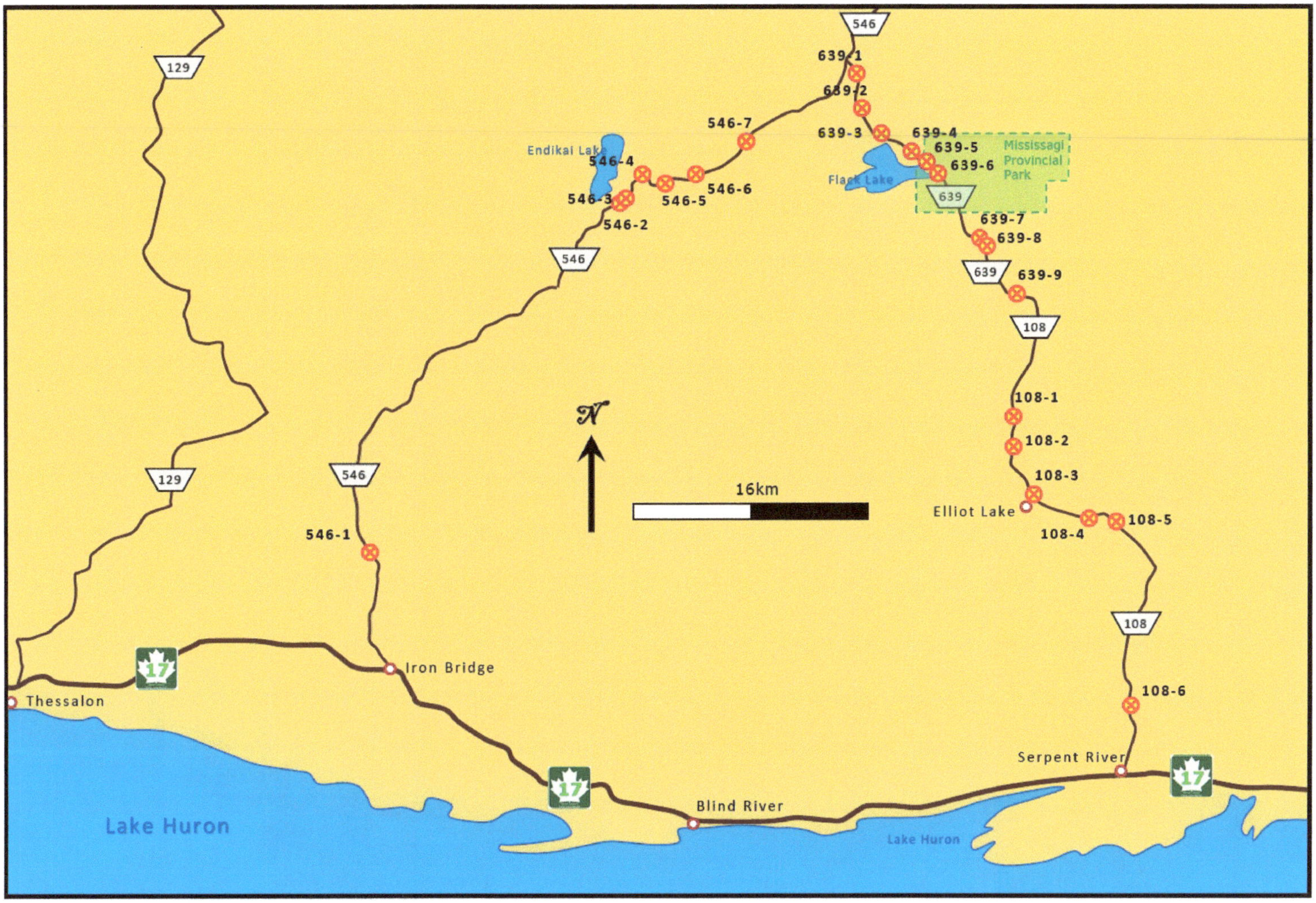

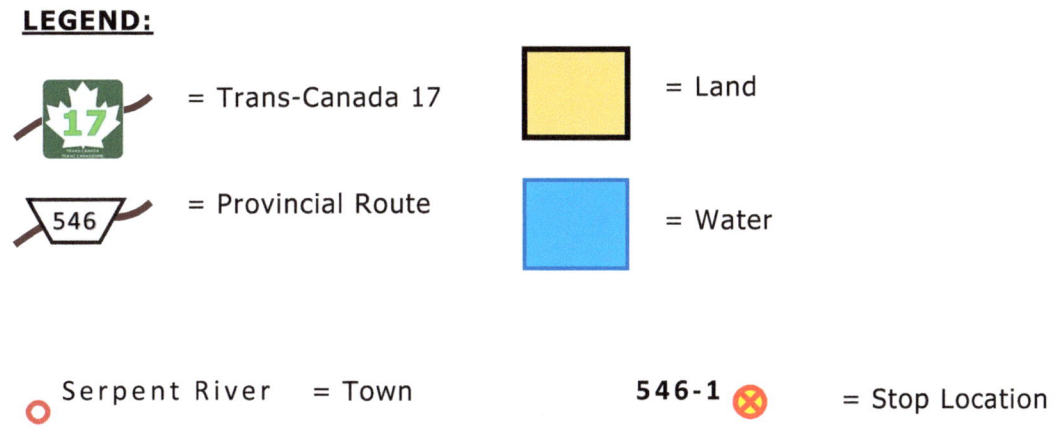

*Adapted from Google Earth (2018)

NOTE: *This map is meant to be a general guide so you can see how the stops relate overall. It is NOT intended to be a detailed map! Only select things are shown (see legend below). Not every body of water is present. Coastlines are somewhat stylized. No rivers or islands are shown.*

LEGEND:

- = Trans-Canada 17
- = Provincial Route
- = Land
- = Water
- Serpent River = Town
- 546-1 ⊗ = Stop Location

Regional Roadside Book Maps (Ontario Only)

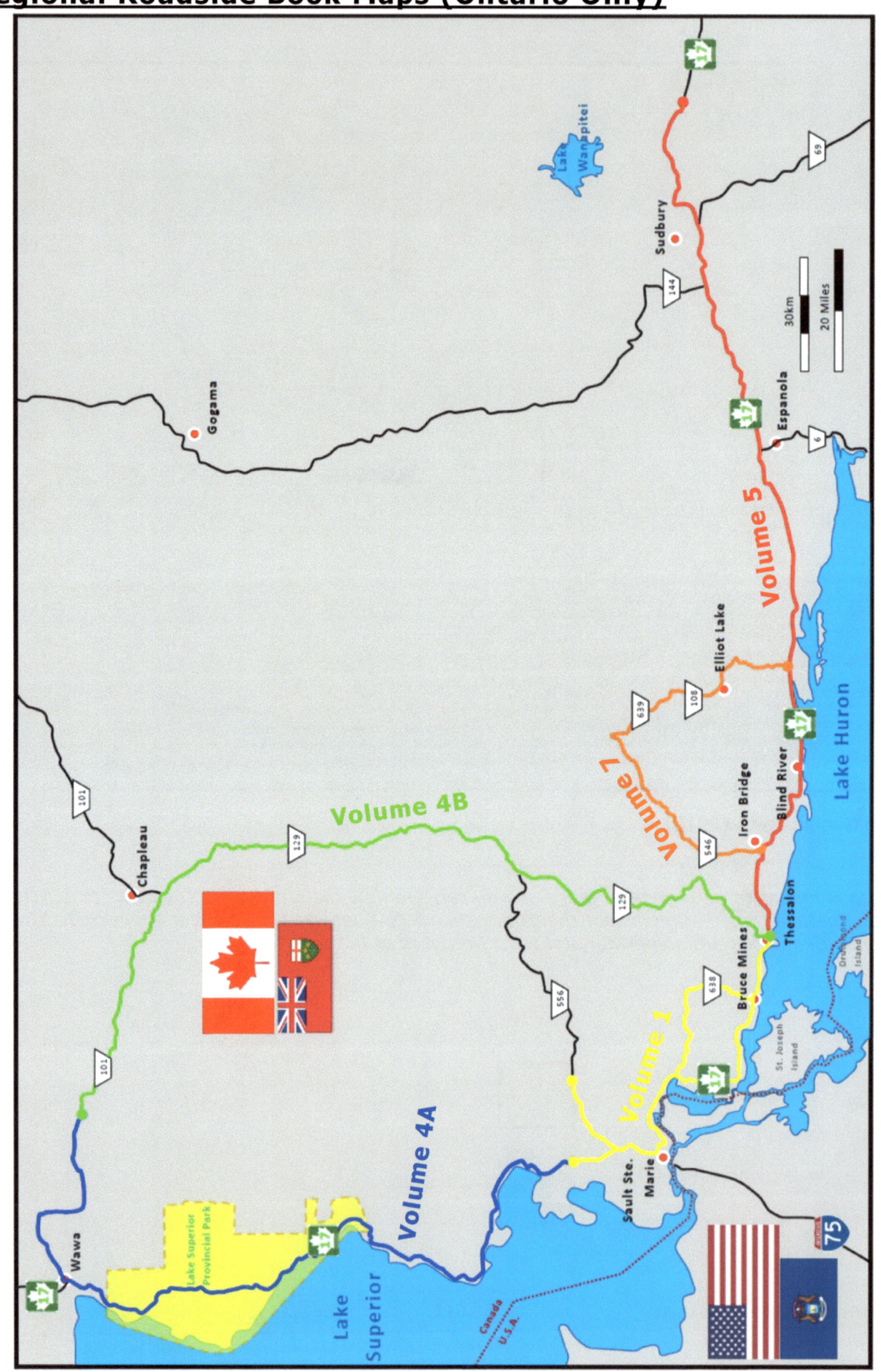

Fieldtrip Stops

Photo was taken by Steven D.J. Baumann on November 29, 2019. See stop 108-2 (p. 58-59).

OUTCROP NAME: 420 meters East-northeast of Eaket Lake Outcrop   **OUTCROP DESIGNATION:** 546-1

OUTCROP LOCATION: GPS: 46.34847 –83.23865   ELEVATION: 258.8m above MSL

FORMAL GEOLOGIC NAME: Gowganda Formation, Coleman Member

MAIN ROCK TYPE(S): Meta-diamictite (tillite), slate, proto-quartzite

DESCRIPTION: This outcrop is on the east side of ON-546. There are a lot of Gowganda outcrops in the area, but this one shows several things. Here you have meta-diamictite that was deposited directly by glaciers as sandy till. The clasts are almost all red and are mostly granitic rocks. There are some proto-quartzite, jasper, quartz, and rare black mafic clasts. The matrix is generally medium to coarse immature quartzite.

Near the north end of the outcrop, interbedded within the meta-diamictite are reddish beds of proto-quartzite and black, laminated slate. The black slate likely represents a shallow lake that existed beneath a melting glacier. The waters were calm, as indicated by the drop stones within it. This unit is almost exactly 1m thick, with consistent thickness throughout the outcrop.

Below the black slate is another meta-diamictite that is underlain by a dark gray, bedded, immature quartzite, with occasional red proto-quartzite. This immature quartzite is not consistently thick and pinches out to the south. Unlike the calm lake deposit above, this unit does not contain drop stones, indicating flowing water. It likely represents a sub-glacial river deposit.

Structurally, the area is simple (see geologic map below). At the outcrop the strike and dip is N15W14SW. Strike varies from north-northwest to north-northeast in the Gowganda. Dips steepen as you head north. There are dikes cross cutting the Gowganda. The 1962 Ontario geologic map (M-2012, p. 70) has them mapped as Keweenaw dikes. However, the 2003, M-2670 map (p. 70) has them mapped as Nipissing intrusions, which is likely correct as they are throughout the area. Many of the maps in the early 1960s to late 1970s often just call any intrusive dikes "Keweenaw", as they had not been dated at the time. We will visit the Coleman Member one more time at stop 108-1 (p. 56-57).

FIGURE: Bedrock geologic Map

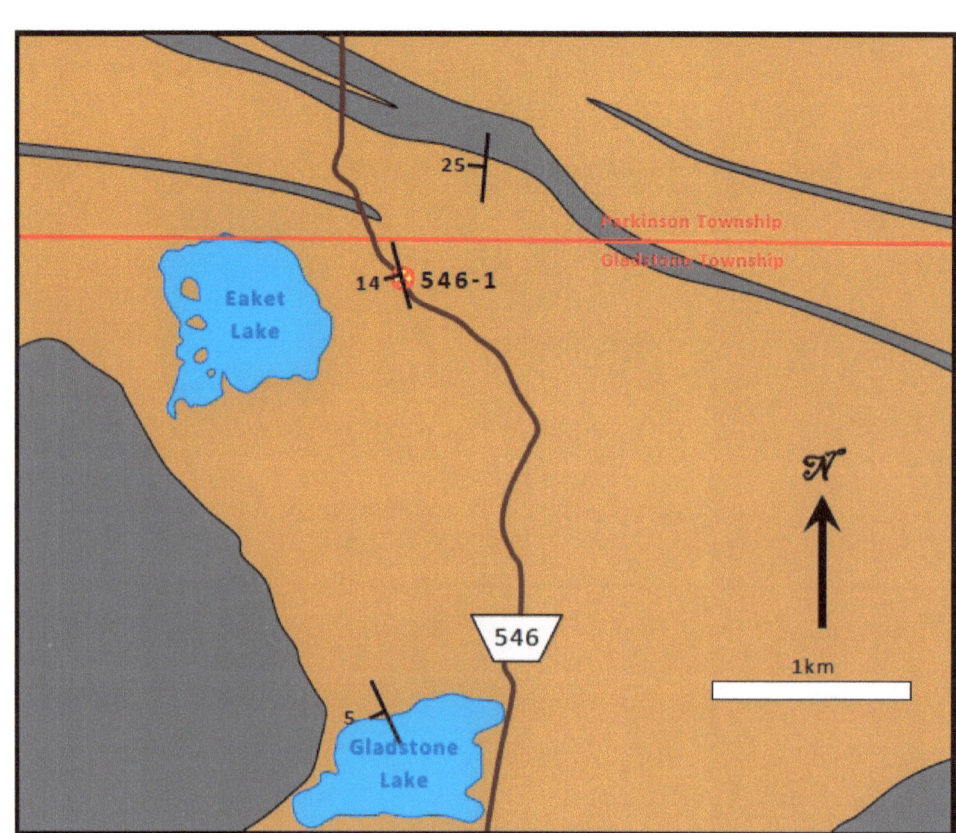

Adapted from Ontario geologic maps M-2012 and M-2670

PHOTOS:

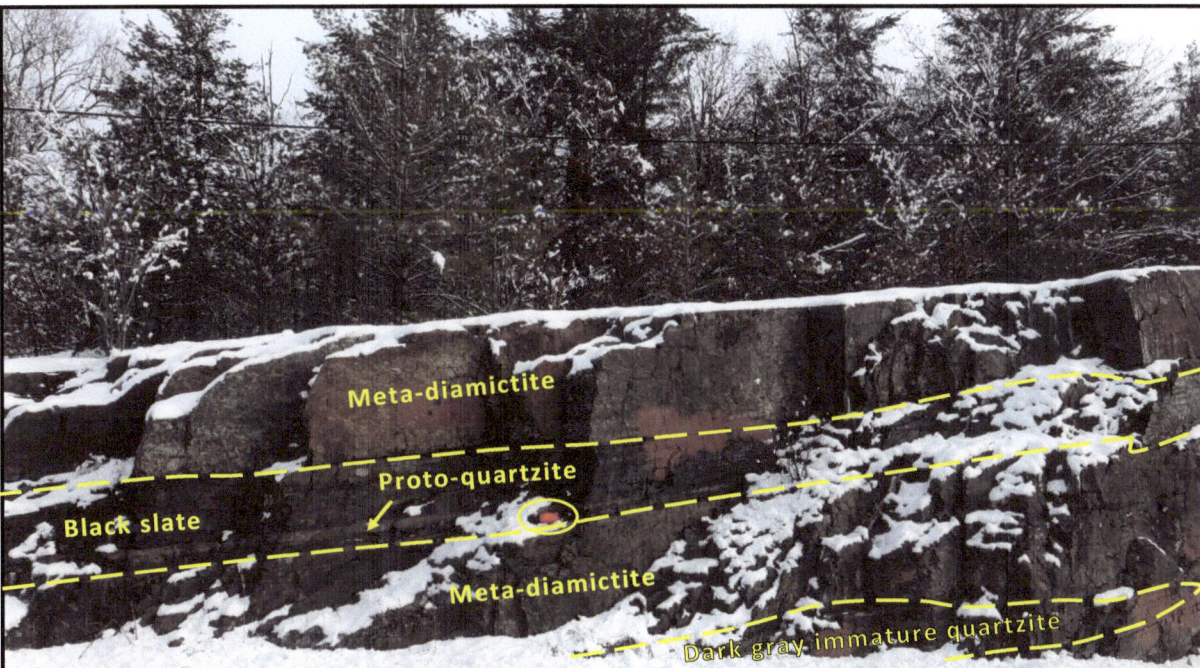

Outcrop looking east-northeast. The interbedded units in the diamictite are labeled. 25cm orange pouch (circled) for scale.

Outcrop looking east-northeast. The proto-quartzite lens (near the base but within the black slate) is outlined in this photo. 25cm orange pouch for scale.

Close-up of the meta-diamictite, looking east-northeast. $2 coin scale.

Close-up of the contact between the black slate and the meta-diamictite, looking east. Yellow arrow is pointing to a drop stone. $2 coin scale.

OUTCROP NAME: Little White River Northeast Narrow Pass Outcrop **OUTCROP DESIGNATION:** 546-2

OUTCROP LOCATION: GPS: 46.56793 −83.01181 **ELEVATION:** 277.4m above MSL

FORMAL GEOLOGIC NAME: Espanola Formation

MAIN ROCK TYPE(S): Meta-siltstone

DESCRIPTION: This outcrop is nestled tightly between the Little White River as it forms a vertical cliff on the east side of route 546.

The rock here is mostly a dark gray meta-siltstone with some light brown laminations. It is slightly calcareous in places and was likely deposited on a shallow continental shelf.

The trend of the rock (which is hard to measure due to the bounding faults) is about N55E25NW, but steepens quickly to nearly a 70° dip near the fault. An unnamed fault parallels the river here , just to the west. Just to the east is the LeSarbo Lake Fault (see bedrock geologic map below). The rocks along this segment of the Little White River are structurally very complex. The largest fault locally (and regionally) is the Flack Lake Fault. The exposed rocks were once deep inside mountains that formed during the Penokean Orogeny ~1895 to 1822Ma and the faults were reactivated during the Midcontinent Rift (~1100Ma) and during the Grenville orogeny to the east ~980Ma.

In the area there has been heavy faulting. The faults were repeatedly reactivated, in either a reverse or strike-slip direction. Due to the multiple orogenic compressional events, shearing has also occurred. At this particular spot, the only obvious deformation (other than tilting) are foliated light brown lenses within the Espanola.

FIGURE: Bedrock geologic Map

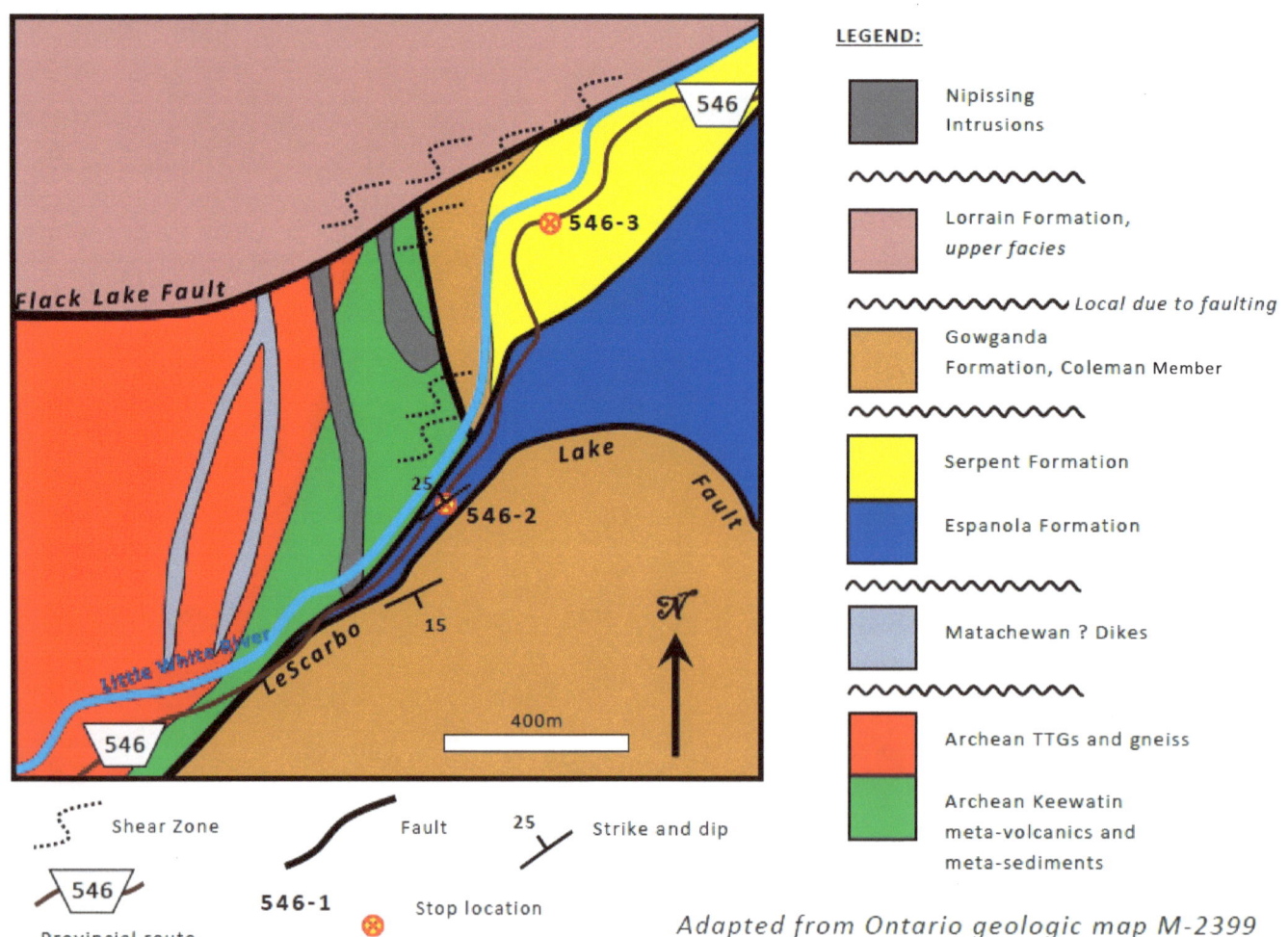

Adapted from Ontario geologic map M-2399

PHOTOS:

Outcrop looking north-northeast. 25cm orange pouch for scale.

Outcrop looking southeast, showing the typical lithology of the Espanola. $2 coin for scale.

Outcrop looking southeast, showing the foliation created by deformation in the light brown lenses. $2 coin for scale.

OUTCROP NAME: Little White River Rapids Pass Outcrop

OUTCROP DESIGNATION: 546-3

OUTCROP LOCATION: GPS: 46.57367 −83.00890

ELEVATION: 280.7m above MSL

FORMAL GEOLOGIC NAME: Serpent Formation

MAIN ROCK TYPE(S): Proto-quartzite

DESCRIPTION: This outcrop is present on both sides of route 546, but is best exposed on the south side of the road.

The unweathered rock here is dominantly a medium crystalline, gray proto-quartzite. There is a slight blue to green hue in places, due to the presence of chlorite and other trace minerals. A gray color in most sedimentary rocks (excluding BIF), usually indicates the presence of organic compounds, but not in this case. The gray color in the Serpent Formation is local and is likely due to the presence of unoxidized iron. This can be hypothesized as indicated by the modern brown rusty weathered surface, which is a weathering of the iron minerals in the Serpent. Unoxidized iron minerals would be expected in the unweathered Serpent Formation because we didn't have free O_2 in the atmosphere. The Serpent Formation is pre-GOE (see p. 6).

This outcrop sits on the south side of one of the largest structures in the region, the flack lake fault. The dominant trend of the Serpent is N66E67SE, but there is some variability. The trend can shallow and turn north in places to about N40E43SE.

FIGURE: Bedrock geologic Map

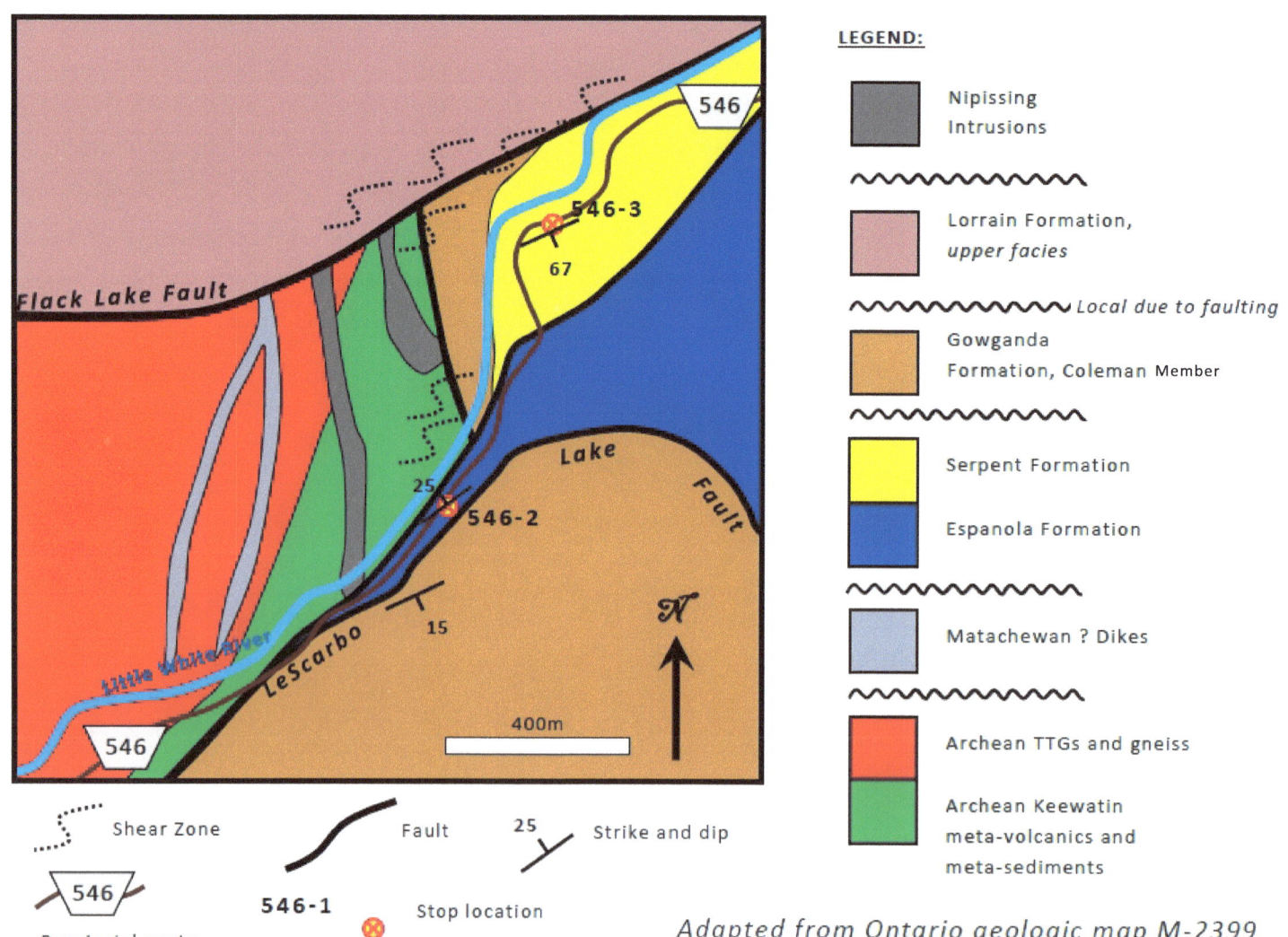

Adapted from Ontario geologic map M-2399

PHOTOS:

Outcrop looking southeast, at the weathered (brown colors) of the Serpent Formation. No scale, but outcrop is about 3 meters tall.

Close-up of the outcrop, looking south at the gray proto-quartzite. $2 coin for scale.

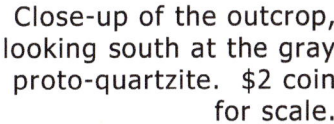

White quartzite vein, looking south. $1 coin for scale.

OUTCROP NAME: North of Little White River Bridge Outcrop **OUTCROP DESIGNATION:** 546-4

OUTCROP LOCATION: GPS: 46.57981 –82.99279 **ELEVATION:** 285.3m above MSL

FORMAL GEOLOGIC NAME: Espanola Formation

MAIN ROCK TYPE(S): Meta-siltstone

DESCRIPTION: This outcrop is located to the north of the Little White River bridge, before the road curves sharply east. There is plenty of parking around here. This area is part of the Little White River Provincial Park, which is one of the many narrow parks that follow rivers in Ontario. This one begins where the Little White River meets the Mississagi River, then extends for about 100km along the Little White to the Blue Lake headwaters (ontarioparks.com/park/littlewhiteriver).

The rocks here are mostly a laminates argillaceous meta-siltstone that varies in color from dark gray to light gray and light brown. The colors follow bedding. This is the middle part of the formation. The lower part is mostly meta-dolostone. The top is locally thin proto-quartzite.

This outcrop is in between two structures. The regional Flack Lake Fault to the north and the Little White River Anticline to the south. The Espanola here trends N55E33SE.

FIGURE: Bedrock geologic Map

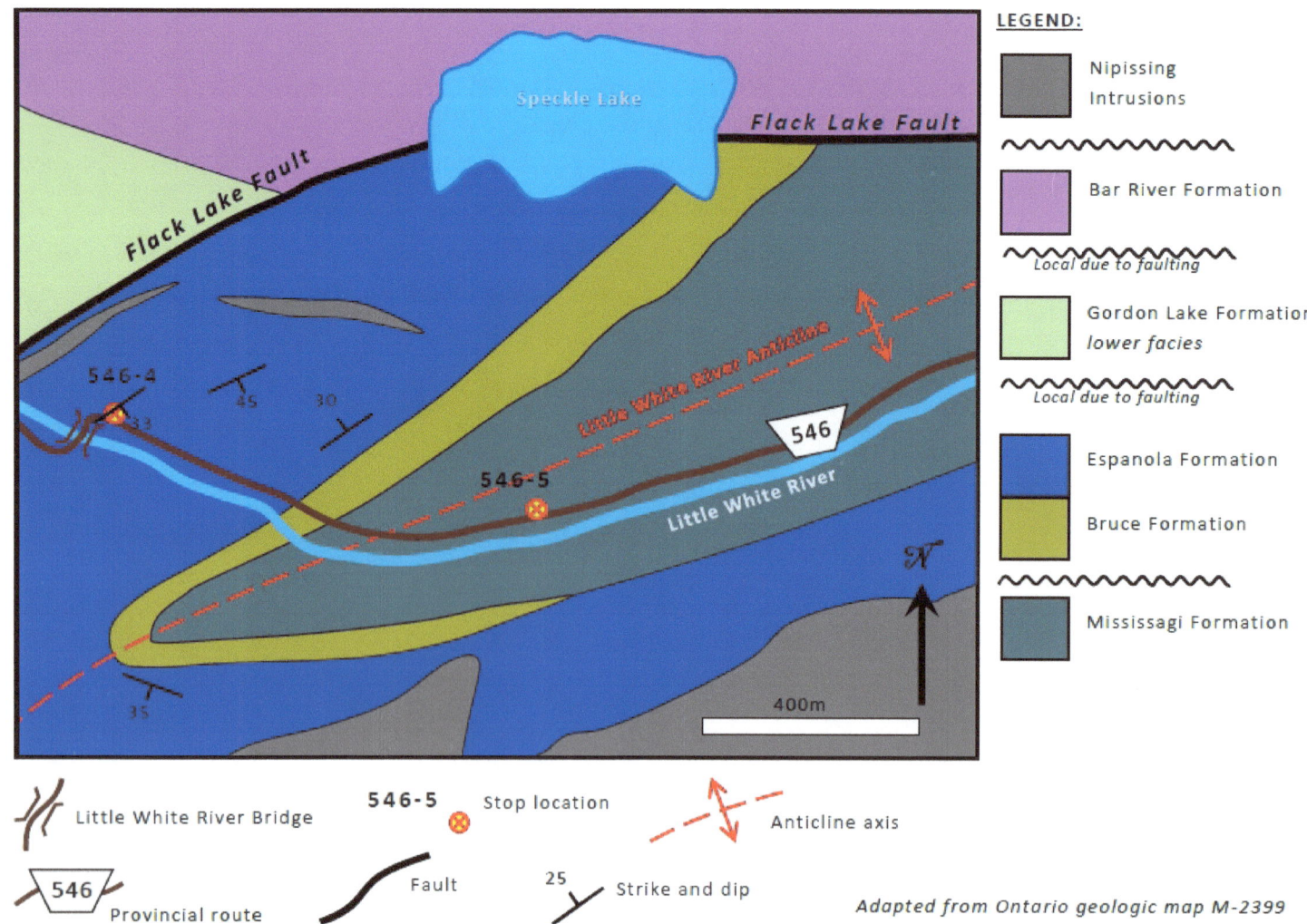

Adapted from Ontario geologic map M-2399

PHOTOS:

The bridge over the Little White River, looking south-southwest from near the outcrop.

Outcrop looking northeast at the Espanola Formation. 25cm orange pouch for scale.

Close-up of the laminations in the Espanola Formation, looking northwest. $2 coin for scale.

OUTCROP NAME: 590m south of Speckle Lake on ON-546 Outcrop   **OUTCROP DESIGNATION:** 546-5

OUTCROP LOCATION: GPS: 46.57817 –82.98170   ELEVATION: 281.9m above MSL

FORMAL GEOLOGIC NAME: Mississagi Formation

MAIN ROCK TYPE(S): Proto-quartzite

DESCRIPTION: This outcrop is located on the north side of route 546. I won't get into great detail on the lithology of the rock here, because we will visit the Mississagi Formation again (stops 639-9 on p. 54-55 and stop 108-3 on p. 60-61). The rock here is a pale yellowish light gray, medium to coarse crystalline, proto-quartzite.

More importantly is the structure. In addition to the bedrock map below, I created a cross section from A-A' on the next page. The cross section gives you the third dimension. A geologic map is a planar surface on an x and Y coordinate plane. The cross section gives you that Z axis and allows you to see inside the Earth. Geologic cross sections can be generated several ways. Shallow ones with good exposure (like here), are generally drawn from surface mapping. Boring logs, cores, seismic maps, gravity anomaly maps, and magnetic maps, can also be used along with surface mapping.

FIGURE: Bedrock geologic Map

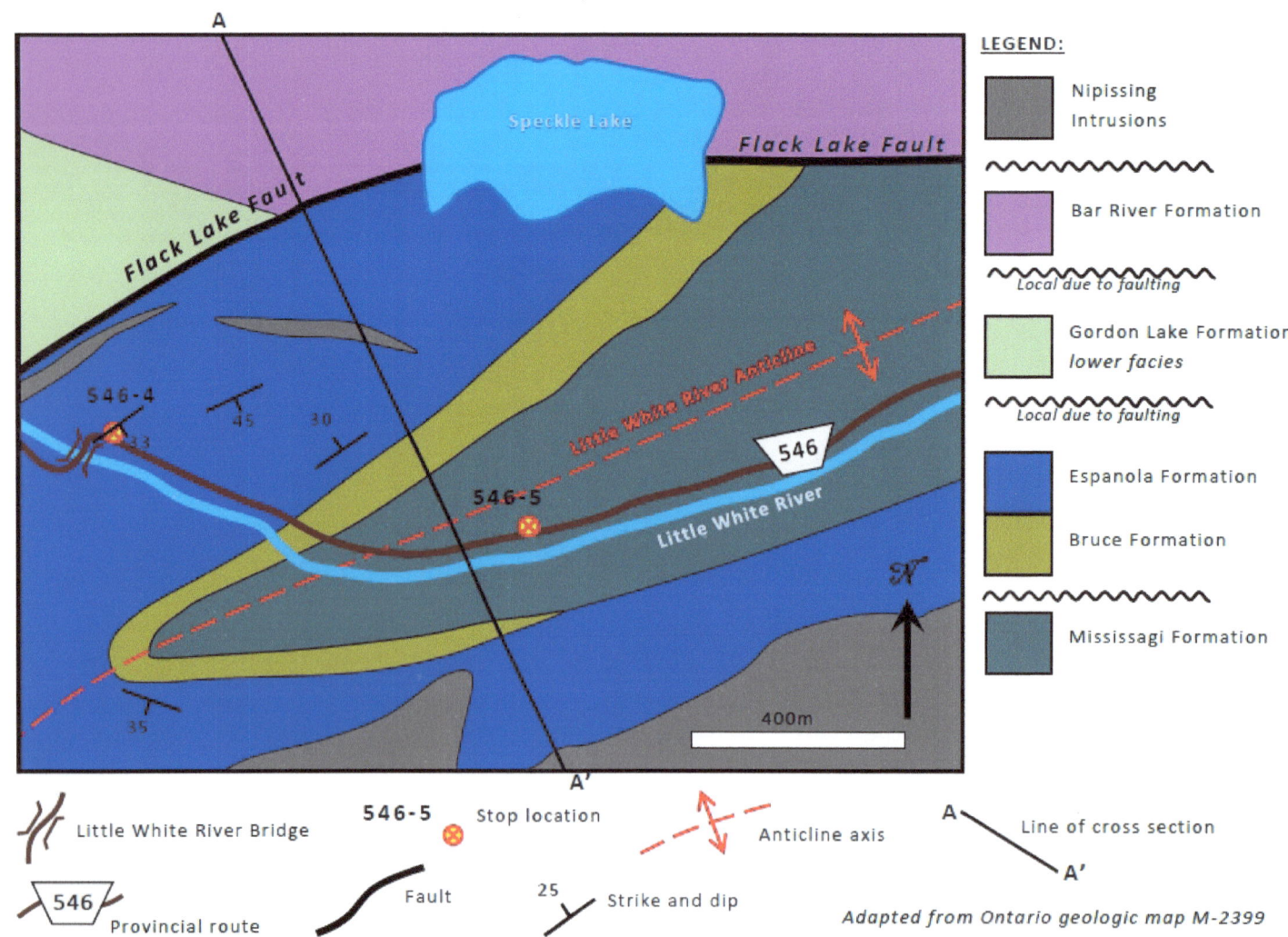

Adapted from Ontario geologic map M-2399

FIGURE: Cross Section A-A'

Cross section from the map on the previous page. Horizontal scale = vertical scale. The colors correspond to the legend on the previous page.

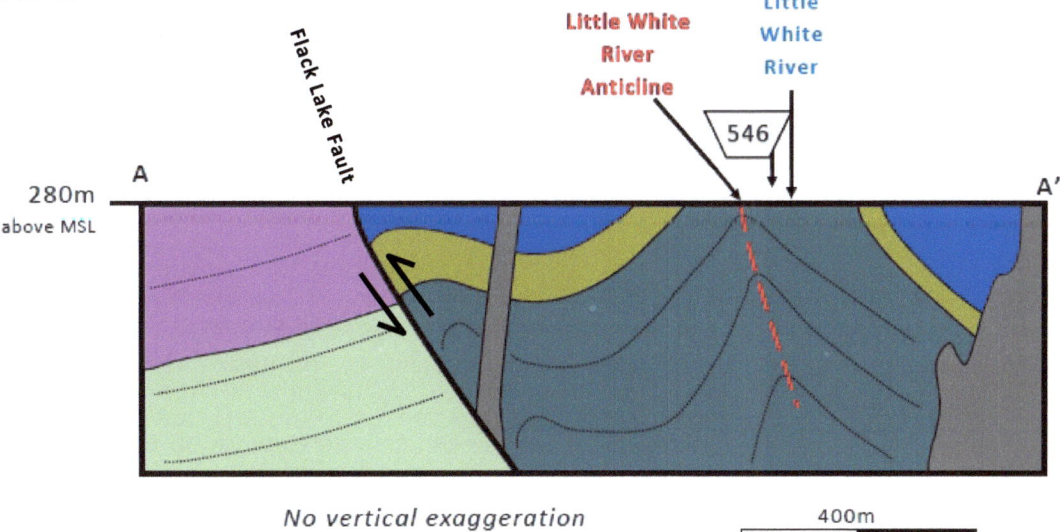

No vertical exaggeration

PHOTOS:

Outcrop, looking west-northwest. No scale but it is about 3 to 5 meters high.

Close-up of the proto-quartzite of the Mississagi Formation, looking north. $2 coin for scale.

OUTCROP NAME: Bend on little white River North Outcrop **OUTCROP DESIGNATION:** 546-6

OUTCROP LOCATION: GPS: 46.58215 −82.95430 **ELEVATION:** 288.0m above MSL

FORMAL GEOLOGIC NAME: Bruce Formation

MAIN ROCK TYPE(S): Meta-diamictite

DESCRIPTION: I almost did not include this outcrop for several reasons. One being, it's difficult to park. Although it is hard to access, it is unfortunately, the best outcrop of the basal Bruce Formation along this route that is exposed nearly all year. We will visit the upper Bruce at stop 108-2, p. 58-59.

The second reason I almost did not include it was because of a mapping error. I did not generate a geologic map due to conflicting information. The outcrop is barely within M-2347, which labels the outcrop as Gowganda Formation. Yet the adjoining map to the west (M-2399) has it as Bruce Formation. Both maps describe the outcrop similarly, but I believe M-2347 is a simple error. It is mapped in between the Mississagi and Espanola Formations, as the Bruce Formation is, yet someone colored it incorrectly and labeled it as Gowganda. The Gowganda is also very thick, on the order of more than 1000 meters, yet M-2347 has it at about 50-100m thick.

Even with the error caught, the two maps do not line up exactly, which is another reason I didn't make a geo map as I didn't have the time to walk around the area. Plus, everything was covered in snow.

Overall the rock is black. It is the metamorphic versions of lithic sandstone to mudstone, with the occasional red granitic clast. There is foliation following bedding, indicating that the rock has been turned nearly vertical. Dip is steep and generally to the north but it varies from 70° to 80° to the north (and may be overturned further uphill), with the dominant trend being about N70W75NE.

FIGURE: Site Location Map

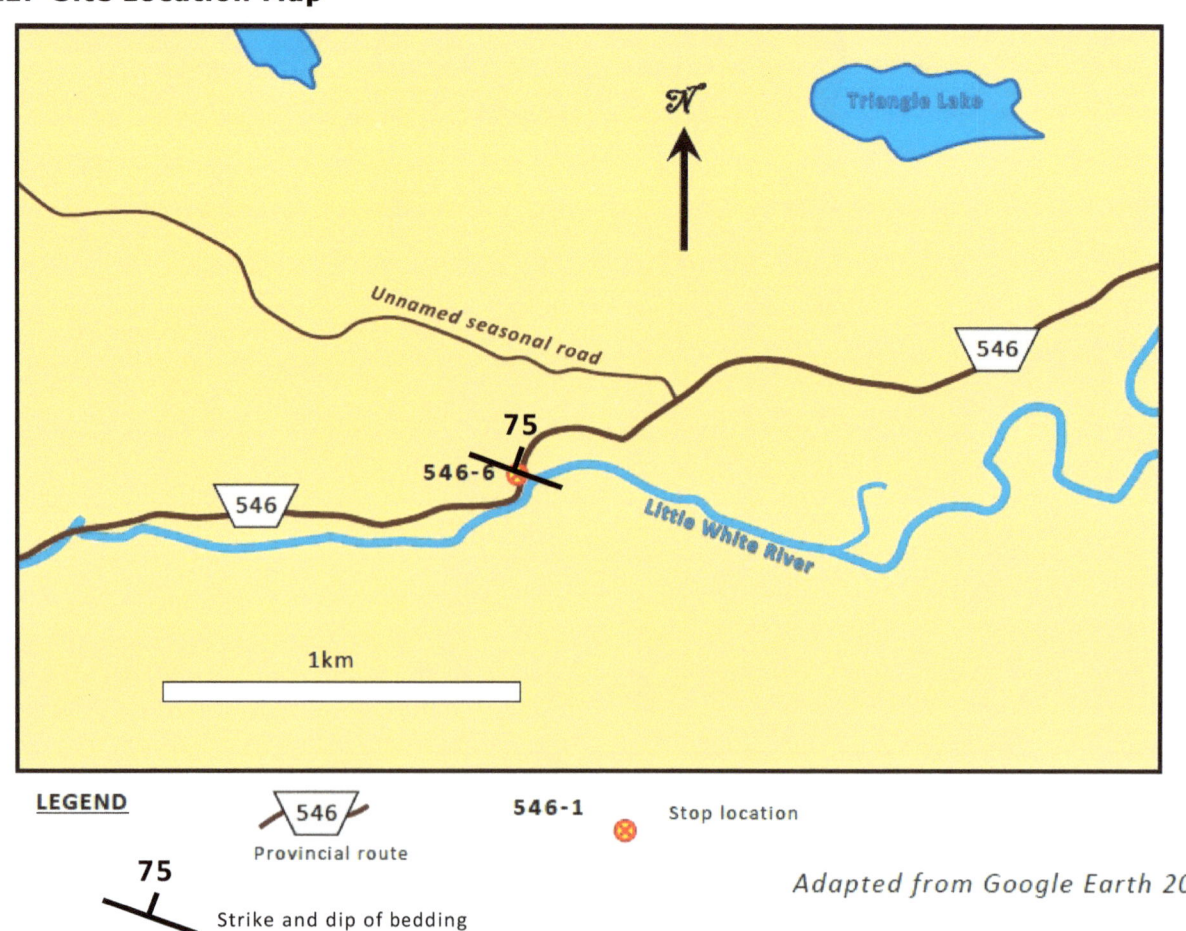

Adapted from Google Earth 2018

PHOTOS:

Outcrop, looking west. No scale but it is about 2 meters high. The yellow dashed lines are the foliations along bedding.

Outcrop, looking west. The yellow dashed line is following a foliation along bedding. The yellow arrows are pointing to red granitic pebbles-cobbles in the meta-diamictite.

OUTCROP NAME: Cliff 1700 meters South-southwest of Bloger's Lake Outcrop

OUTCROP DESIGNATION: 546-7

OUTCROP LOCATION: GPS: 46.60285 −82.90664

ELEVATION: 296.6m above MSL

FORMAL GEOLOGIC NAME: Lorrain Formation (middle facies)

MAIN ROCK TYPE(S): Jasper conglomerite (pudding stone)

DESCRIPTION: The Lorrain in the area is extensively exposed in the hills to the west of ON-546. Structurally, the area is simple. The rock trends west-east (more or less) and dips <40° to the south. All the bedding strike and dips on the geologic map below are from Ontario M-2347. I did not take strike and dip. The closest rocks to the road were in large boulders. It was to icy and slippery for me to navigate over the talus to reach the cliff, which is about 60 meters from the road (see photos on the next page).

The Lorrain Formation dominates the area. Here the middle facies is well exposed. We will visit the upper facies later. The middle facies is the most beautiful and sought after part of the formation. It is dominated by a white pebble conglomerite that contains abundant red jasper. This is commonly called "pudding stone". It is called this because there are clasts (in this case mostly red jasper with minor red arkose and slate) that contrasts sharply with the surrounding rock (in this case white quartz pebbles and quartzite), thus resembling raisin/Christmas pudding. The middle Lorrain is the coarsest of the 3 facies and is often massive to cross bedded.

The red clasts can range from scattered (like in the boulder on the next page), or they can take up more than 50% of the rock. The depositional environment is terrestrial. The middle Lorrain was deposited in fluvial braided streams with slight initial dip.

FIGURE: Bedrock geologic Map

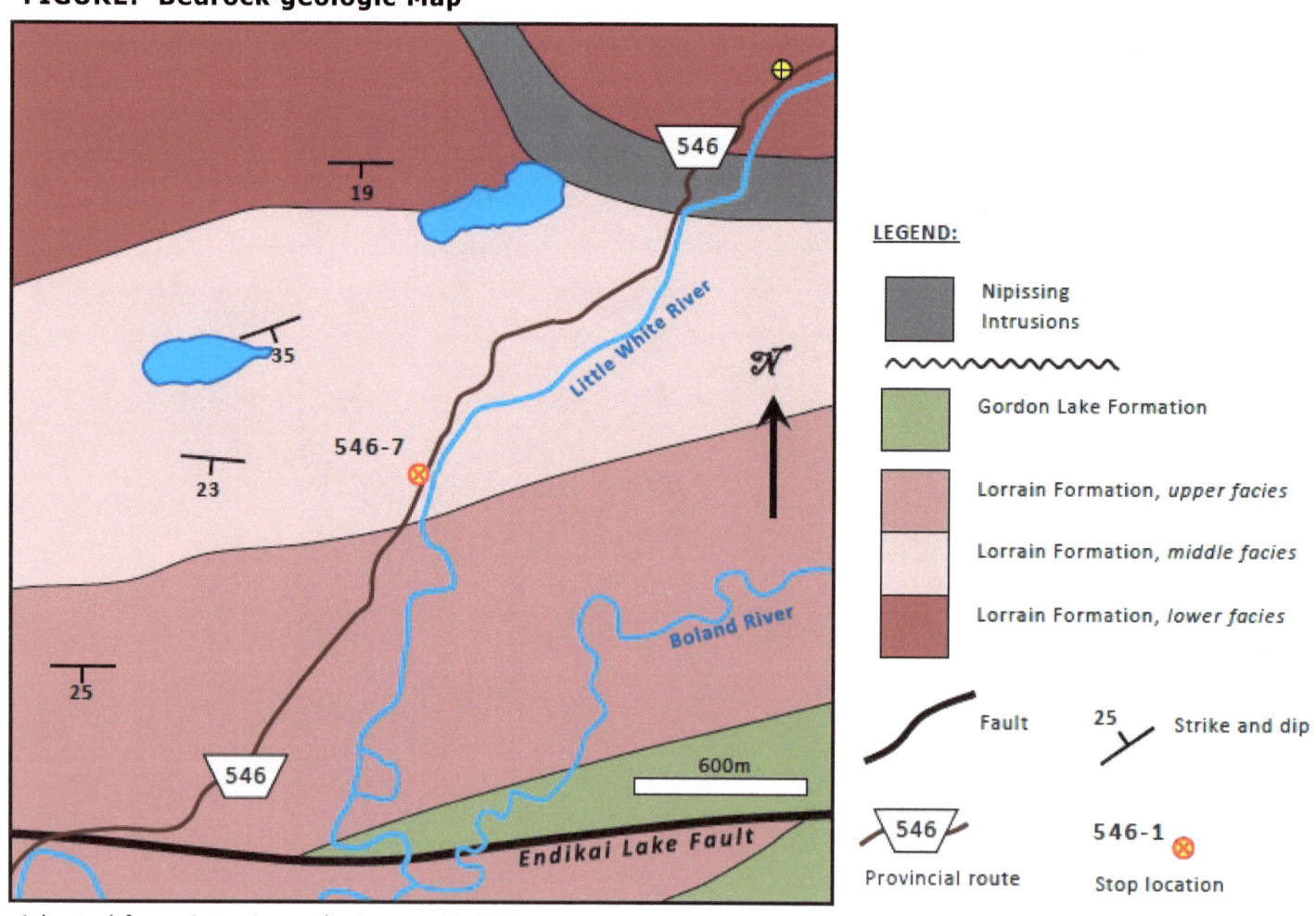

Adapted from Ontario geologic map M-2347

PHOTOS:

Outcrop, looking west-northwest. No scale but the tall cliff in the background is about 130 meters higher than road level and is set back about 60 meters from the road.

Talus boulder looking north-northeast. $2 coin scale.

Zoom-in of the photo A. The red clasts in photo A are the same ones in photo B. Here you see the typical lithology of the middle facies of the Lorrain Formation.

OUTCROP NAME: 455 meters Northwest of Cobre Lake on ON-639 Outcrop

OUTCROP DESIGNATION: 639-1

OUTCROP LOCATION: GPS: 46.64293 −82.80610

ELEVATION: 331.3m above MSL

FORMAL GEOLOGIC NAME: Lorrain Formation (upper facies)

MAIN ROCK TYPE(S): Quartzite

DESCRIPTION: Here we turn off of route 546 and begin to head back south towards the town of Elliot Lake. We will also be leaving the Little White River behind, from this point on.

Here the rock is the upper facies of the Lorrain. It is a nearly white fine-medium crystalline quartzite. It likely was deposited as fluvial braided stream deposits, but closer to the coast. It lacks the coarse clasts that dominate the underlying middle facies. The bedding here is general medium level to medium cross bedded, within larger bed sets (see photos on the next page). Bedding can be hard to pick out on fresh surfaces.

Although strike and dip wasn't taken, the rock has an apparent east-west trend with an apparent dip of <7° to the south (see photos on the next page). Actual strike is likely more northeast with actual dip being more southeast and between 8° and 10°, as shown by the other strike and dips within the middle facies (see geologic map below).

The geologic map below only contains the strike and dips of the local dominant trend. Some variation in the strike and dip of the Lorrain is present near the Nipissing intrusions. The below map is also somewhat simplified. There are numerous suspected small faults in the map area, which are all left out here because it would have made the map too busy. Since they aren't key to understanding the local let alone regional geology, they were left off.

FIGURE: Bedrock geologic Map

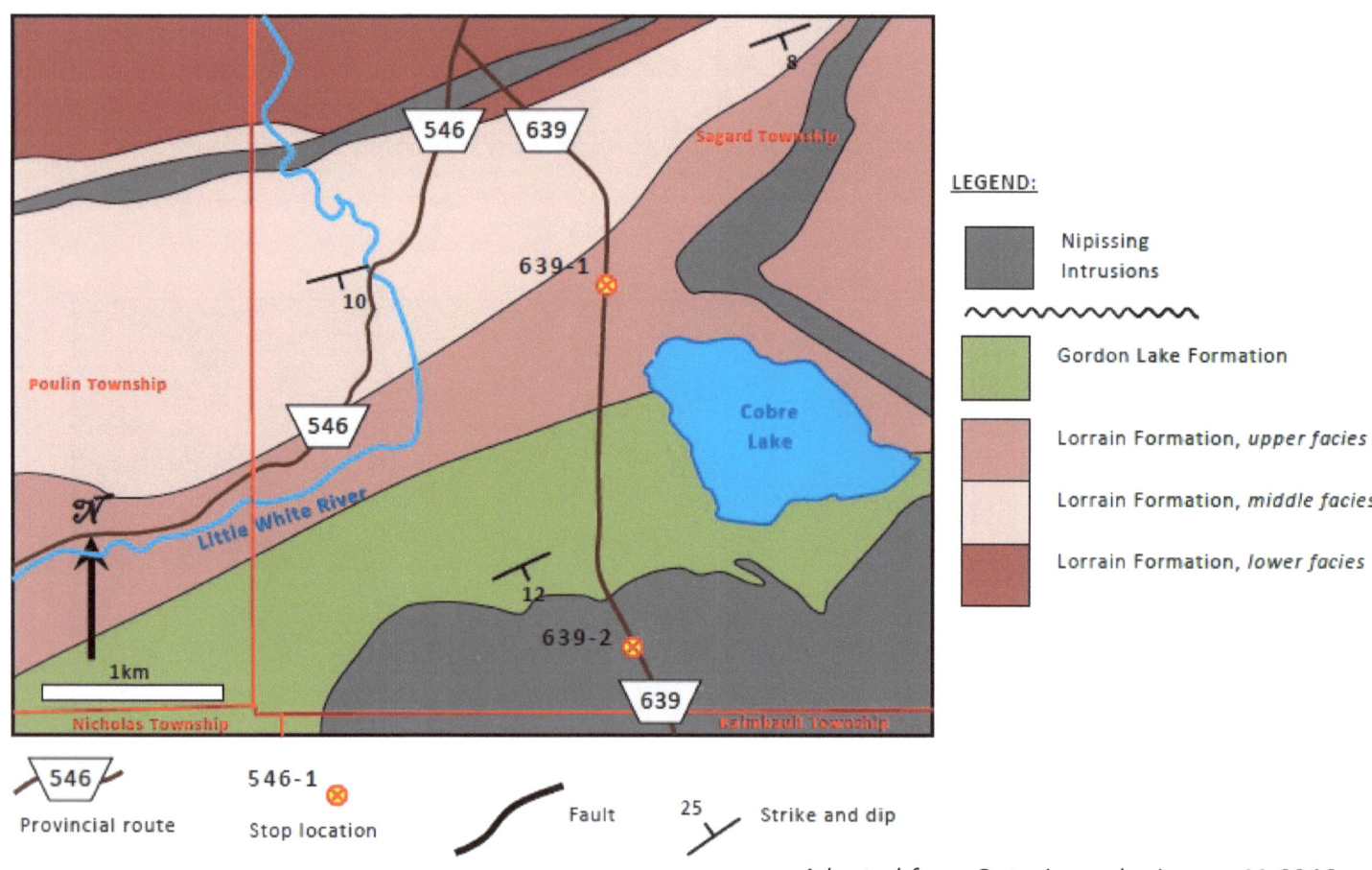

Adapted from Ontario geologic map M-2346

PHOTOS:

Long outcrop, looking north-northeast. Truck for scale.

Looking east at the upper facies. 5'10" (1.78m) person for scale. The black lines highlight the dominant bedding and bed sets. The dashed lines mark some of the cross beds.

Close-up of a fresh break in the quartzite, looking east. $1 coin for scale. Notice the lack of obvious bedding.

OUTCROP NAME: 1057 meters Southwest of Cobre Lake on ON-639 Outcrop

OUTCROP DESIGNATION: 639-2

OUTCROP LOCATION: GPS: 46.62414 –82.80449

ELEVATION: 446.5m above MSL

FORMAL GEOLOGIC NAME: Nipissing intrusions

MAIN ROCK TYPE(S): Monzogabbro to quartz monzogabbro

DESCRIPTION: This outcrop is actually several low lying outcrops on both sides of the road (see photos on the next page).

The Nipissing cross cuts all of the geologic units in the Huronian Supergroup and it is younger than all the other formations in the Huronian. The Nipissing is actually a series of intrusions with slightly different ages and lithologies. The Nipissing intrusions vary from diorite to gabbro to basalt.

The Nipissing was generally deposited as intrusion from upwelling magma. Mostly as dikes and sills. However, this location represents a small pluton. For whatever reason, the Nipissing here stopped as it intruded the Lorrain Formation and it cooled slowly underground. We know this because of the coarse crystals in the rock.

The Nipissing intrusions are somewhat of an enigma as their emplacement does not correspond to any local orogeny. We are pretty confident that the ocean basin that grew and allowed for the deposition of the Huronian; ultimately closed with the Penokean Orogeny 1895Ma to 1822Ma (see p. 14). It's possible they were intruded as magma when the passive Huronian margin became an active area of subduction that ultimately take ~300 million years for the ocean basin to close, thus completing one of the earliest Wilson Cycles.

FIGURE: Bedrock geologic Map

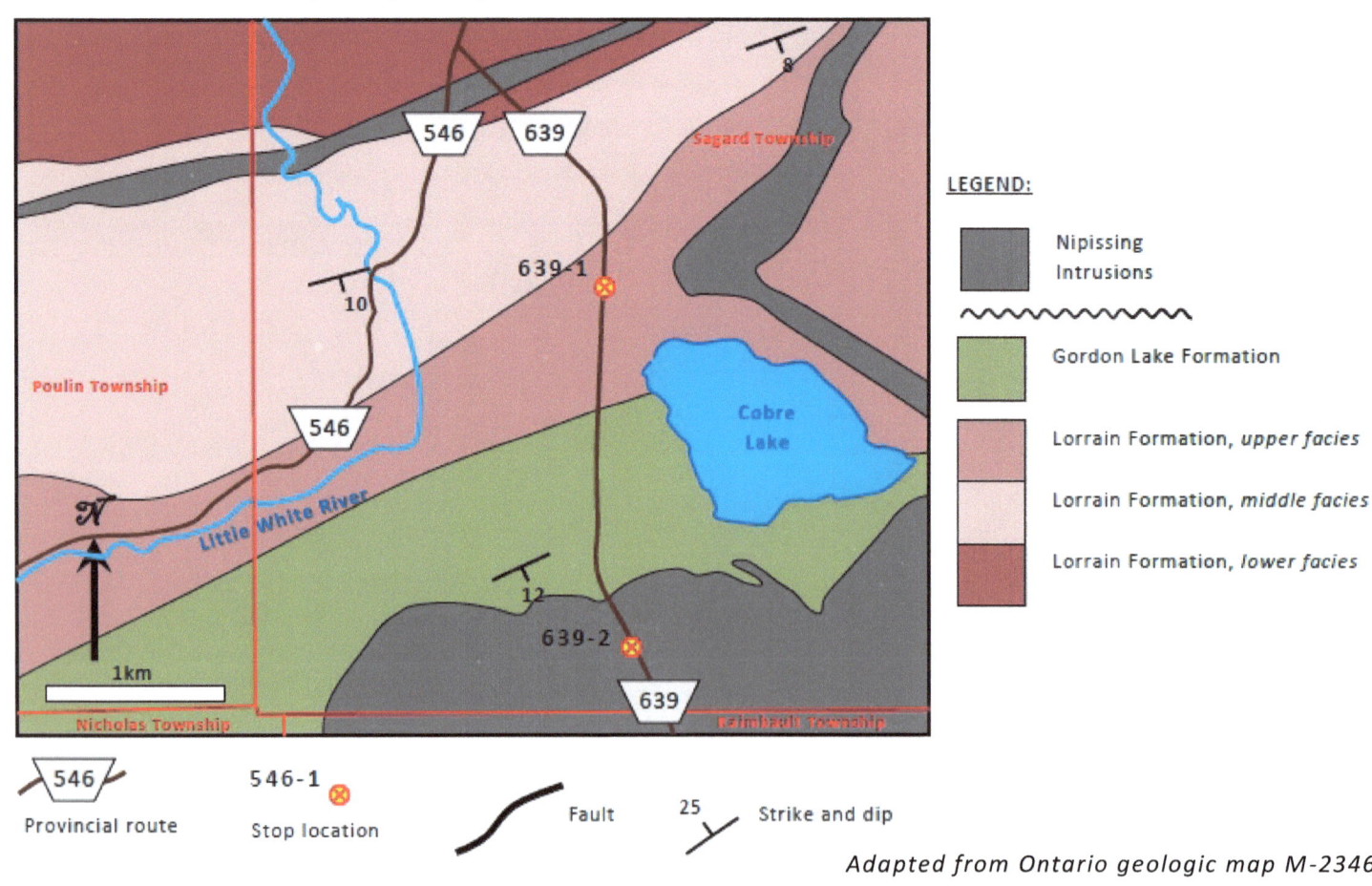

Adapted from Ontario geologic map M-2346

PHOTOS:

The low lying Nipissing outcrops on both sides of route 639, looking south-southeast. No scale. Flack Lake is visible in the background.

Close-up of the plagioclase (light colored part of the rock) and dark mineral (pyroxenes) in the gabbro, looking southwest. $2 coin scale.

OUTCROP NAME: 208 meters South of Boland River/510 meters North of Flack Lake Outcrop

OUTCROP DESIGNATION: 639-3

OUTCROP LOCATION: GPS: 46.60817 −82.78799

ELEVATION: 340.2m above MSL

FORMAL GEOLOGIC NAME: Gordon Lake Formation

MAIN ROCK TYPE(S): Siltite and argillite

DESCRIPTION: This is unfortunately the only outcrop of the Gordon Lake formation that we will be visiting. I say unfortunate, because off the main roads is where vermiform fossils have been found in abundance within the Gordon Lake. Vermiform structures have been controversial for a long time since their first documentation back in the 1960's. At first they were thought to be multi-cellular, then they were thought to not be fossils at all, and now they are back on the fossil page but as colonial microorganisms, like stromatolites. The Huronian is generally devoid of fossils but the Cobalt Group (the group the Gordon Lake is a formation in the Cobalt Group) has yielded vermiforms (Hill et al., 2016).

Locally the Gordon Lake Formation is a metamorphosed siltstone (siltite) and shale (argillite). There is minor chert and proto-quartzite, but not at this outcrop. Here the Gordon Lake laminated and gray on fresh surfaces. It weathers a distinct olive color with some rust on the surface (see photos on the next page). As you head west towards Sault Ste. Marie, the Gordon Lake begins to become more complex. It was likely deposited in tidal flat to nearshore marine deposits.

Here the rock trends about N75E10SE. The dip here is shallow and like stops 639-1 (p. 38-39) and 639-2 (p. 40-41), the structure is simple.

FIGURE: Bedrock geologic Map

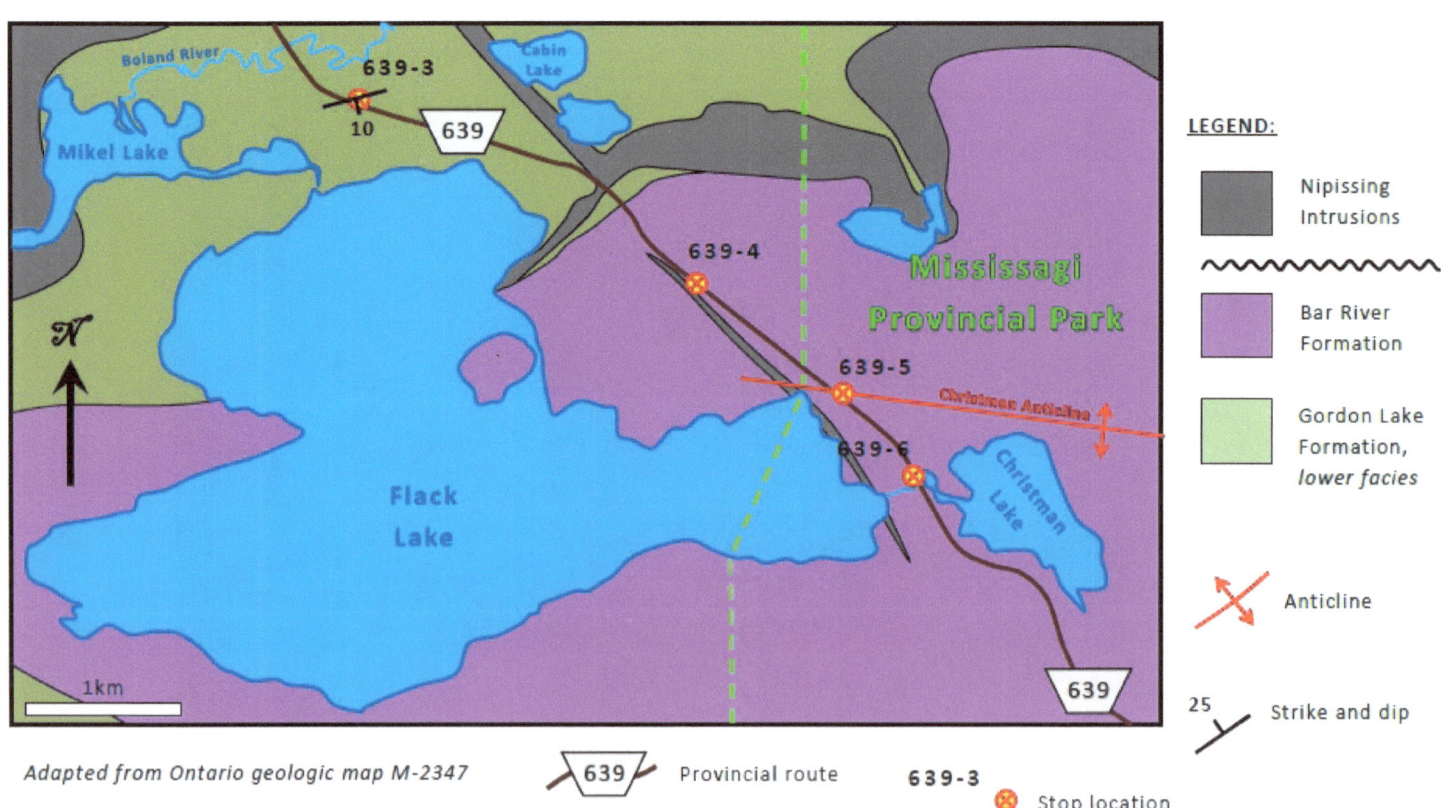

PHOTOS:

Outcrop looking west-northwest at the upper facies. 5'10" (1.78m) person for scale.

Laminated siltite and argillite, looking south-southeast. $2 coin scale.

OUTCROP NAME: 150 meters West of No Name Lake Outcrop

OUTCROP DESIGNATION: 639-4

OUTCROP LOCATION: GPS: 46.59783 –82.75909

ELEVATION: 413.9m above MSL

FORMAL GEOLOGIC NAME: Bar River Formation

MAIN ROCK TYPE(S): Quartzite

DESCRIPTION: The next three stops are going to be of the Bar River Formation, all showing different things.

This outcrop is the typical lithology of the Bar River Formation. It is a white fine to medium crystalline quartzite. It lithologically is very similar to the Lorrain Formation, at least at this outcrop. Although both were originally quartz rich sandstone, they are not lithologically identical. The Bar River does not contain any pudding stone. It also tends to contain more hematite than the Lorrain Formation. Although not in the area, the Bar River contains white shale made of Kaolin clay. Something that is absent in the Lorrain. Also unlike the Lorrain, the Bar River cannot be easily divided into facies or members.

The Bar River is the youngest known formation of the Cobalt Group and of the sediments of the Huronian Supergroup. It was likely deposited in a mix of nearshore marine and fluvial deposits, punctuated by small lakes.

Not only is this outcrop typical of the Bar River, but it is also pretty much undeformed. Since it is quartzite it had to be somewhat deeply buried to undergo low grade metamorphism. At one time after the deposition of the Bar River but before the Penokean Orogeny, there had to be at least several kilometers of overlying sediments that are now gone...just eroded away. If those younger sediments still exist, they have yet to be discovered.

FIGURE: Bedrock geologic Map

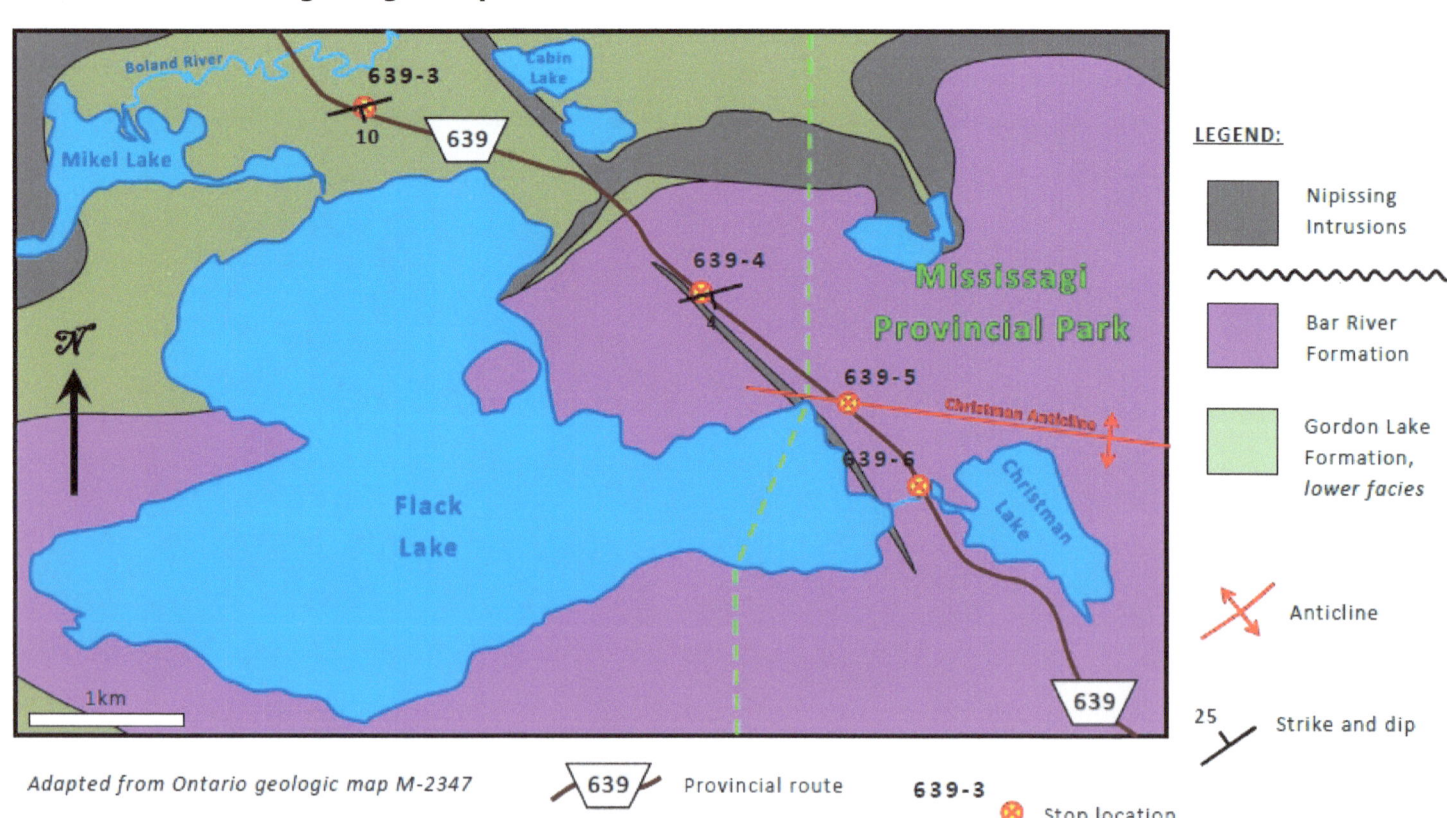

PHOTOS:

Outcrop of the Bar River Formation, looking southwest. Notice the nearly level bedding.

OUTCROP NAME: 103 meters Northeast of Flack Lake Outcrop **OUTCROP DESIGNATION:** 639-5

OUTCROP LOCATION: GPS: 46.59145 −82.74758   ELEVATION: 358.1m above MSL

FORMAL GEOLOGIC NAME: Bar River Formation

MAIN ROCK TYPE(S): Quartzite

DESCRIPTION: This outcrop is within Mississagi Provincial Park, so please do not take samples. Here the rock is essentially identical to the Bar River at stop 639-4, but the lithology isn't why I included this stop.

This stop is right at the crest of an anticline structure (see geologic map below and photos on the next page). It's somewhat extensive although not very dramatic. It was important enough to be included on Ontario's 1977 map M-2347. On that map it extends east-southeast for nearly 4000 meters.

Although we model the axis (the line that follows the highest point of the anticline, or its crest) as straight as possible, it is more irregular than indicated. You can see that even with my two strikes and dips. Even though it isn't a perfect textbook example of a non-plunging anticline, it's close. It's somewhat asymmetrical, varying only 4° between each limb near the axis. Dip on the north limb steepens as you go further into the Earth. As where it begins to level out the further south you go from the axis.

The really cool thing about this outcrop is that the road cut is nearly perpendicular to the axis of the anticline. When I first saw it I knew I had to make sure it actually was an anticline and not an erosional artifact or unloading structure, making it look like an anticline.

FIGURE: Bedrock geologic Map

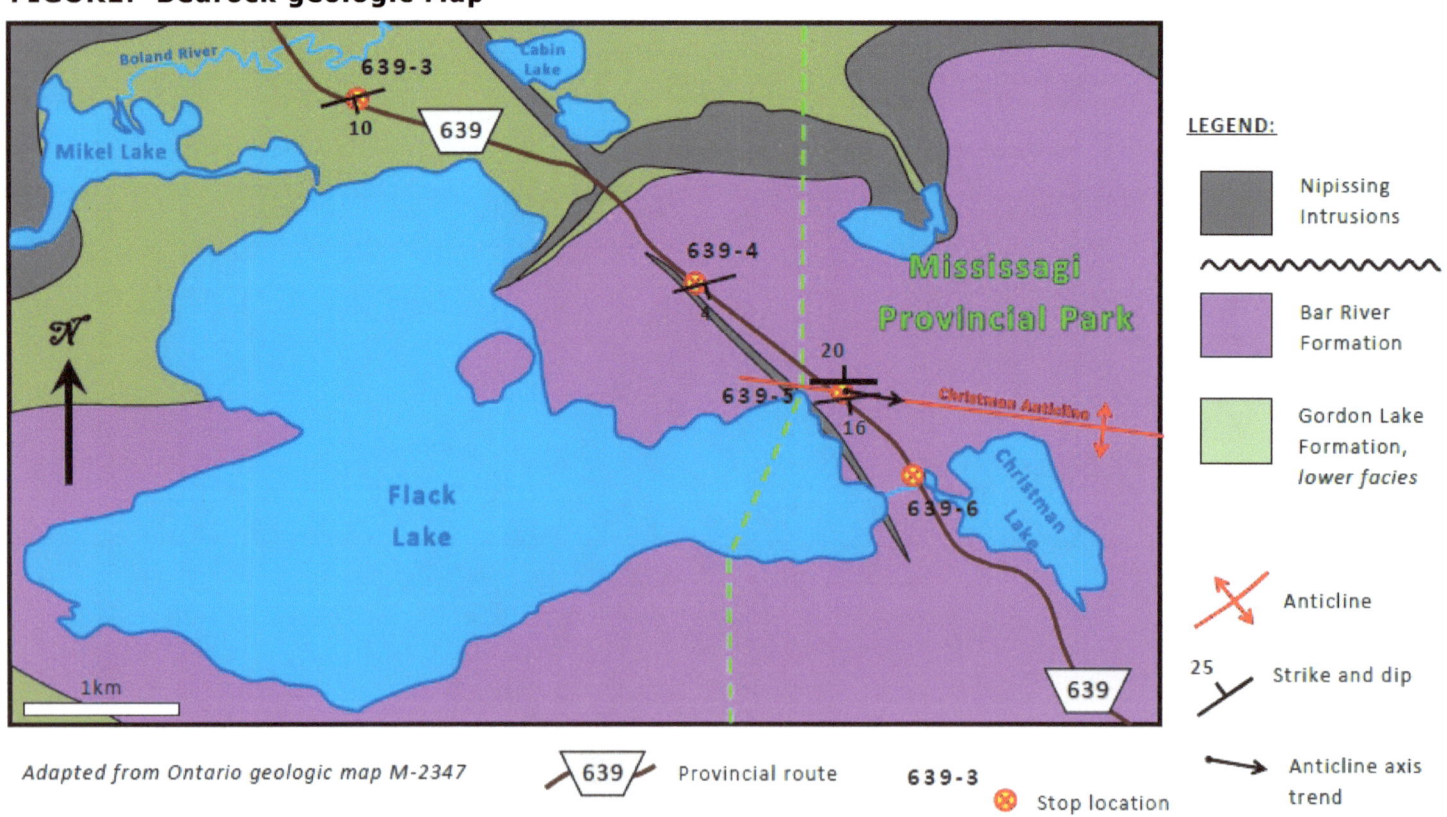

Adapted from Ontario geologic map M-2347

PHOTOS:

Outcrop without all the labels, looking east-southeast. No scale (see below).

Outcrop, looking east-southeast. The yellow dashed lines follow bedding. The red arrow points along the trend direction of the anticline axis and the numbers above it give the trend. The strike and dip on the left corresponds to the graphic to the north of the axis the geologic map (previous page), the right one corresponds to the south strike and dip.

OUTCROP NAME: North of Flack-Christman Lake Connection Outcrop **OUTCROP DESIGNATION:** 639-6

OUTCROP LOCATION: GPS: 46.58621 –82.73957 **ELEVATION:** 360.9m above MSL

FORMAL GEOLOGIC NAME: Bar River Formation

MAIN ROCK TYPE(S): Hematitic quartzite

DESCRIPTION: Here the grains that form the quartzite of the Bar River are the same as those at stops 639-4 (p. 44-45) and 639-5 (p 46-47). The only real difference is the much thinner bedding and the abundant hematite in the rock. There is so much of it in places that it exists as purple laminations between quartzite laminations. The surface color is due to the weathering out of the hematite that exists between the grains.

Near this location (if not this exact one) is also where well developed vermiform fossils were recovered in the 1965 (see photo below).

FIGURE: Bedrock geologic Map

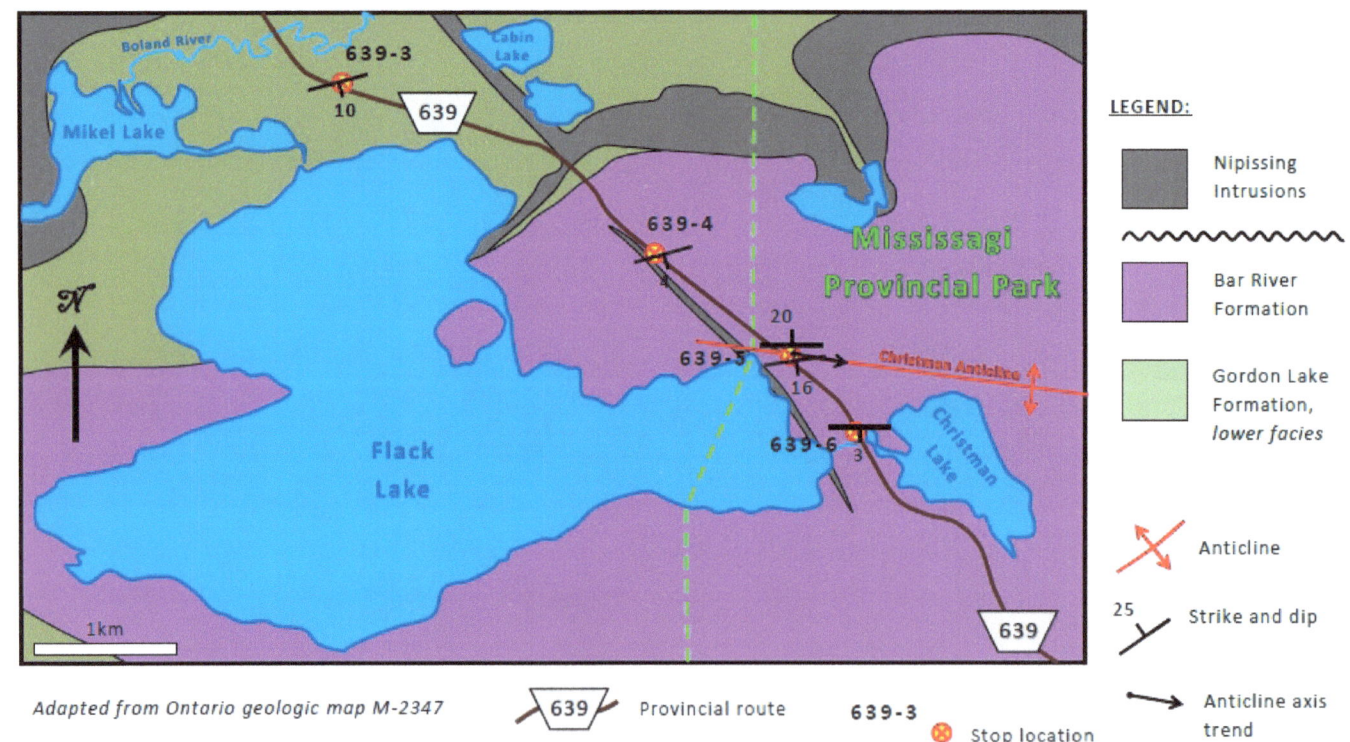

PHOTOS:

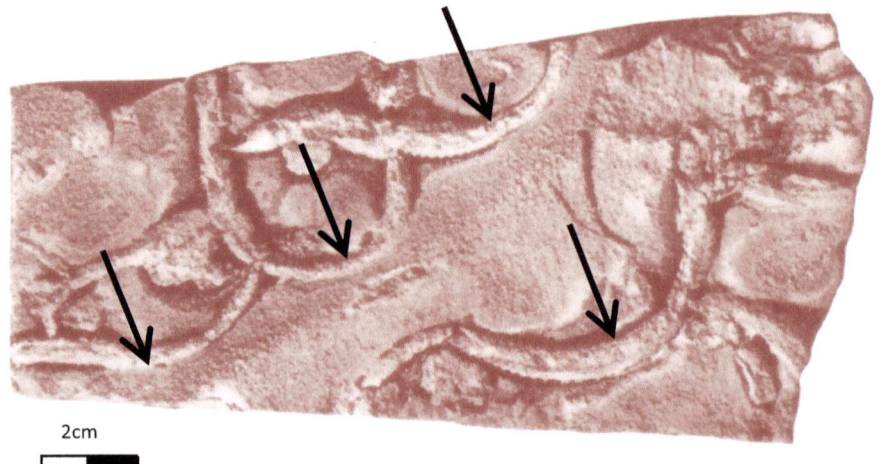

Adapted from Hofmann (1971), plate 13, figure 1. It was collected in 1965 by V. Lahti and named "*Rhysonetron lahtii*". Black arrows are pointing to the vermiform structures that are the fossil. Not in true color.

PHOTOS (continued):

Outcrop, looking northwest, truck for scale.

Outcrop, looking southwest. Hematite coats the surface, giving the rock its red color, but it also exists between the grains. $2 coin scale.

Zoom-in of the pic up and to the left. Here you can see the dark purple bands of hematite that follow bedding (yellow arrows). $2 coin scale.

OUTCROP NAME: 15 meters West of Lake Twenty Three Outcrop   **OUTCROP DESIGNATION:** 639-7

OUTCROP LOCATION: GPS: 46.54449 −82.69913   ELEVATION: 439.5m above MSL

FORMAL GEOLOGIC NAME: Archean TTG

MAIN ROCK TYPE(S): Tonalite-granodiorite

DESCRIPTION: As we head further south and cross the Quirke Lake Fault (see geologic map below) we leave the Huronian Supergroup and we enter the Archean basement rock.

These rocks are the oldest known plutonic rocks in the area. They vary from nearly undeformed (like this outcrop) to migmatite (not visited). Many of these rocks are informally called the "Algoma Suite, Gneiss, Granitoid Complex and Laurentian Granite" (Leith, et al., 1935). Very little has been done to figure out the exact internal relationships of these rocks. Some individual plutons and suites have been subdivided out, but only in areas of economic interest, mostly in the greenstone belts to the west and northwest. Depending on what you consider "Algoma Suite" the oldest igneous felsic (p.19-20) rocks are either 2727Ma or 2697Ma. With the youngest at 2665Ma (Heather, et al., 1995). Despite this inconsistency due to lack of nomenclature stability, the rocks are definitely Neoarchean.

At this outcrop the rock is pretty much in its original crystallization state. There isn't any significant metamorphism. The crystals are easy to see with the unaided eye, and are coarse...almost pegmatitic, Although very red, the dominant feldspar is not K-spar, it is plagioclase. It is a granodiorite to tonalite. Being Archean, it is what we call a TTG. TTG's were very common in the Archean and true granites were rare. Today the opposite is true. That basic observation clues us in to the fact that modern Plate Tectonics has not always been in operation on Earth.

FIGURE: Bedrock geologic Map

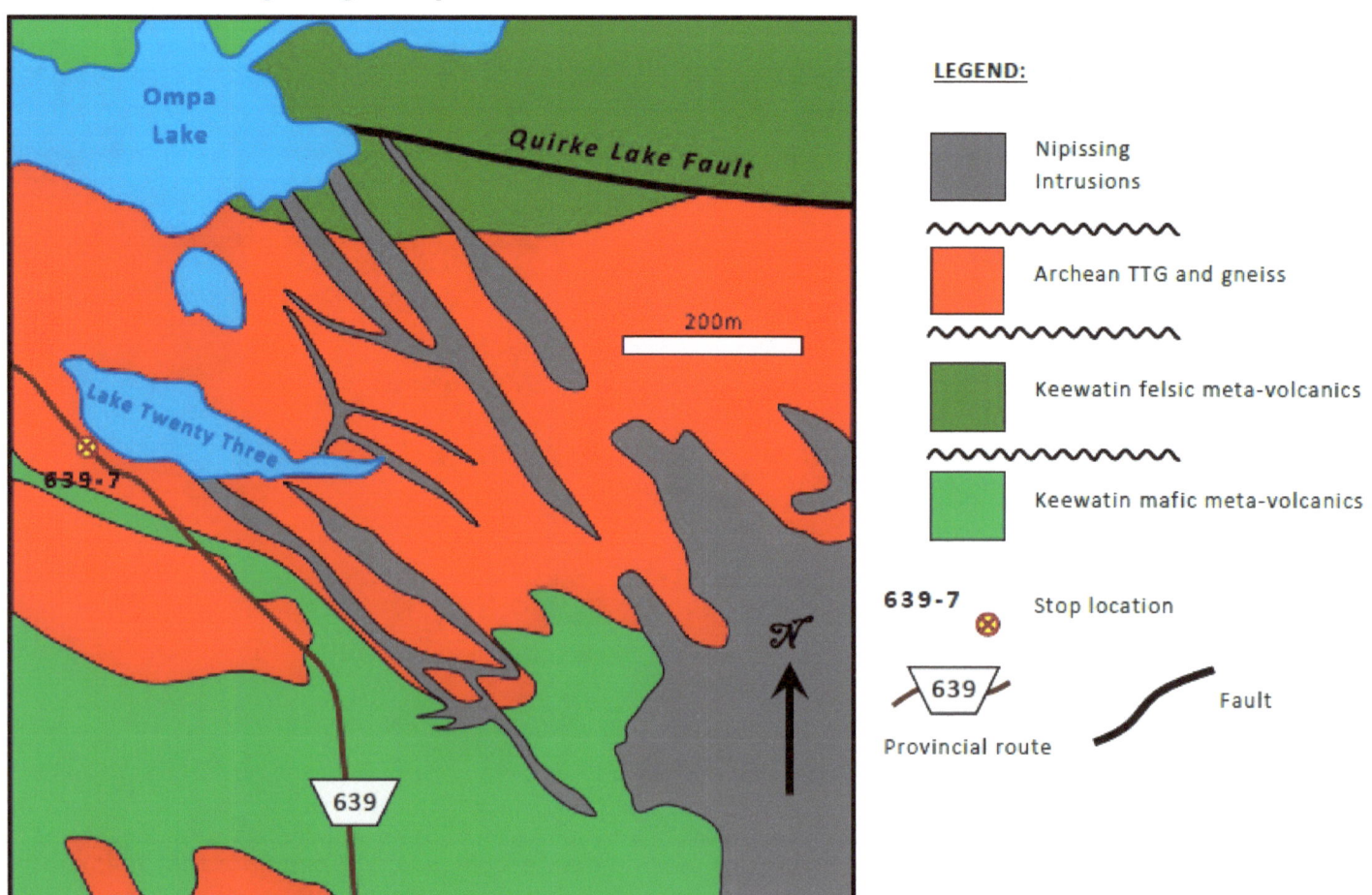

Adapted from Ontario geologic map M-2306

PHOTOS:

Outcrop, looking southwest, truck for scale.

Outcrop, looking southwest. $2 coin scale.

Outcrop, looking southwest. Notice how coarse and undeformed the rock is. $2 coin scale.

OUTCROP NAME: 580 meters South-southeast of Lake Twenty Three Outcrop

OUTCROP DESIGNATION: 639-8

OUTCROP LOCATION: GPS: 46.53859 –82.69121

ELEVATION: 438.6m above MSL

FORMAL GEOLOGIC NAME: Keewatin Series

MAIN ROCK TYPE(S): Meta-basalt

DESCRIPTION: The Keewatin Series is mostly mafic to felsic (p.19-20) meta-volcanics and meta-sedimentary rocks. They are usually nearly black in color and mostly fine crystalline. At this outcrop, they are likely a meta-basalt or a closely related aphanitic rock, like meta-andesite.

They are the oldest in the region. How old? Well, that's a hard question to answer. Based on field relationships, the Keewatin is slightly older than Timiskaming Series (in the Cobalt area of Ontario) and the Algoma rocks intrude both (uwaterloo.ca/earth-sciences-museum/resources/mining-ontario/geological-history-cobalt). The oldest Algoma is about 2727Ma (see stop 639-7), so it has to be older than that. To allow some time for the Timiskaming meta sediments, it is likely older than 2800Ma. No successful dating of the Keewatin has occurred. It's very old, finer grained, and has been deformed in at least 3 orogenies (Algoman, Penokean, and Grenville orogenies). So the rocks are at least 2800Ma, possibly much older. This would make them Mesoarchean (2800-3200Ma) or older.

FIGURE: Bedrock geologic Map

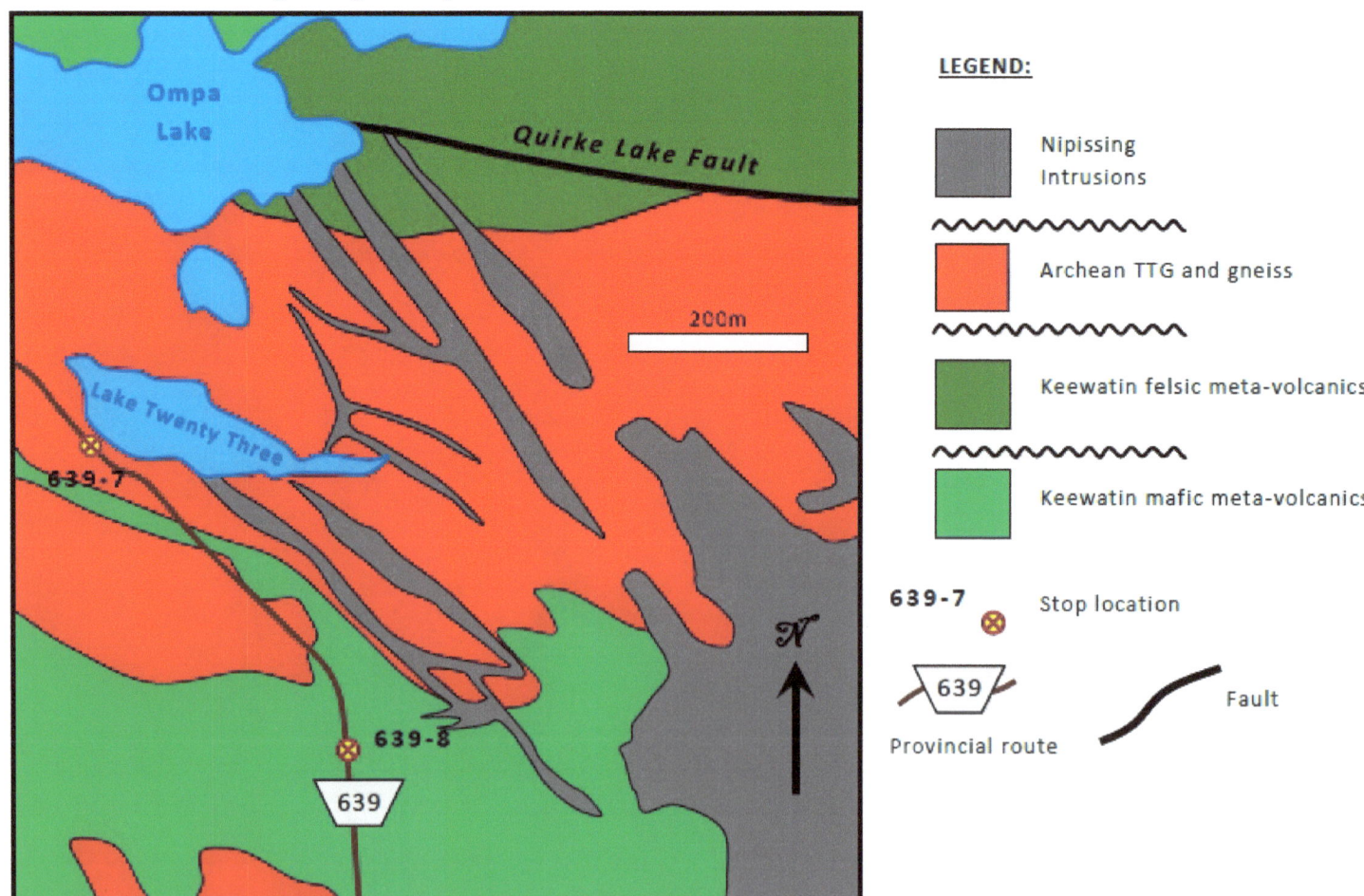

Adapted from Ontario geologic map M-2306

PHOTOS:

Outcrop, looking west-southwest. No scale.

Outcrop, looking west-southwest, The "M15" is about 30cm in height.

OUTCROP NAME: 620 meters East-southeast of ON-639 and Gravelpit Lake Overpass Outcrop

OUTCROP DESIGNATION: 639-9

OUTCROP LOCATION: GPS: 46.51147 –82.66558

ELEVATION: 406.6m above MSL

FORMAL GEOLOGIC NAME: Mississagi Formation (upper facies)

MAIN ROCK TYPE(S): Proto-quartzite to quartzite

DESCRIPTION: At this outcrop we are near the very top of the upper Mississagi Formation. The abandoned Quirke Mine no.1 shaft, is about 1.5km to the east-northeast of the outcrop. The Mississagi is a uranium bearing formation (IAEA-TECDOC-427, 1987) and the it was mined here from 1956 to 1990, temporally closed from 1960 to 1968. The mine had a no.1 and a no.2 shaft.

The Mississagi originated as a sandstone and conglomerate. The pre-GOE sedimentary rocks, like the Mississagi, routinely contain uranium ore. The upper Mississagi is likely a fluvial stream deposit. In a reducing atmosphere, uraninite (uranium ore that use to be called "pitchblende") and other minerals like pyrite are stable. That is how we know there was no free oxygen in the atmosphere when the Mississagi was deposited. If there was any free oxygen in the atmosphere (as the Mississagi was likely a terrestrial deposited) uraninite would have broken down and any pyrite would weather leaving red beds behind.

The Mississagi's distinct yellowish gray color stands out sharply against the surrounding gray rocks. At this particular location is close to the proto-quartzite/ quartzite line. We will be visiting the lower Mississagi later (not locally deposited) at stop 108-3 (p.60-61).

Due to the limited exposure at the time I visited this outcrop, I did not take a strike and dip from the outcrop. On the geologic map below; I have included other nearby strike and dips from the Ontario geologic map M-2114.

FIGURE: Bedrock geologic Map

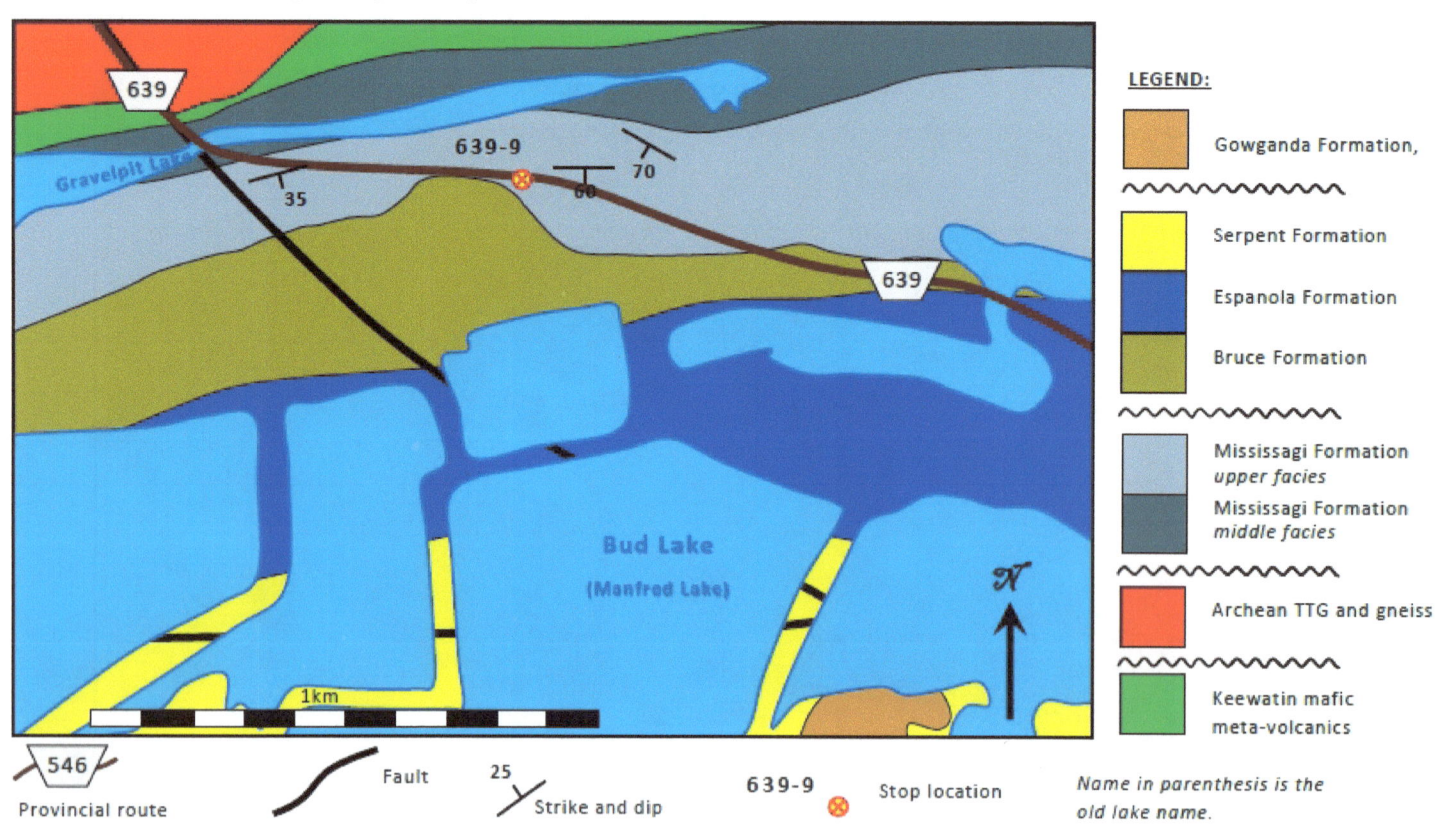

Adapted from Ontario geologic map M-2114

PHOTOS:

Outcrop, looking southwest. 5'10" (1.78m) person for scale.

Outcrop, looking west-southwest. Notice the color. It is somewhat a unique characteristic of the Mississagi Formation in the area covered in this book.

OUTCROP NAME: 380 meters South of Mitchell Road on the West Side of ON-108 Outcrop

OUTCROP DESIGNATION: 108-1

OUTCROP LOCATION: GPS: 46.42994 –82.66804

ELEVATION: 396.8m above MSL

FORMAL GEOLOGIC NAME: Gowganda Formation, Colman Member

MAIN ROCK TYPE(S): Proto-quartzite and conglomerite

DESCRIPTION: Here we see the Coleman Member of the Gowganda Formation again, but this time we are near the bottom of the member.

I really like this outcrop, not for the lithological variation (see photos on the next page), but for the varying clasts in the conglomerite. There aren't only red granitic and proto-quartzite. There's also greenish granitic rocks with pink felsic veins, gray reworked gray immature to proto-quartzite quartzite, rare mud clasts, rare red jasper, and even some dolostone clasts. When a conglomerite has so many variable clasts it is referred to as a polymictic conglomerate.

This is a great outcrop to observe the cycles of glacial deposits. The bottom clast supported conglomerite (or diamictite), are overlain with deep purplish gray pebbly proto-quartzite (occasionally immature quartzite). Both of these are topped with either a sorted mix of proto-quartzite and conglomerite or more polymictic conglomerite.

FIGURE: Bedrock geologic Map

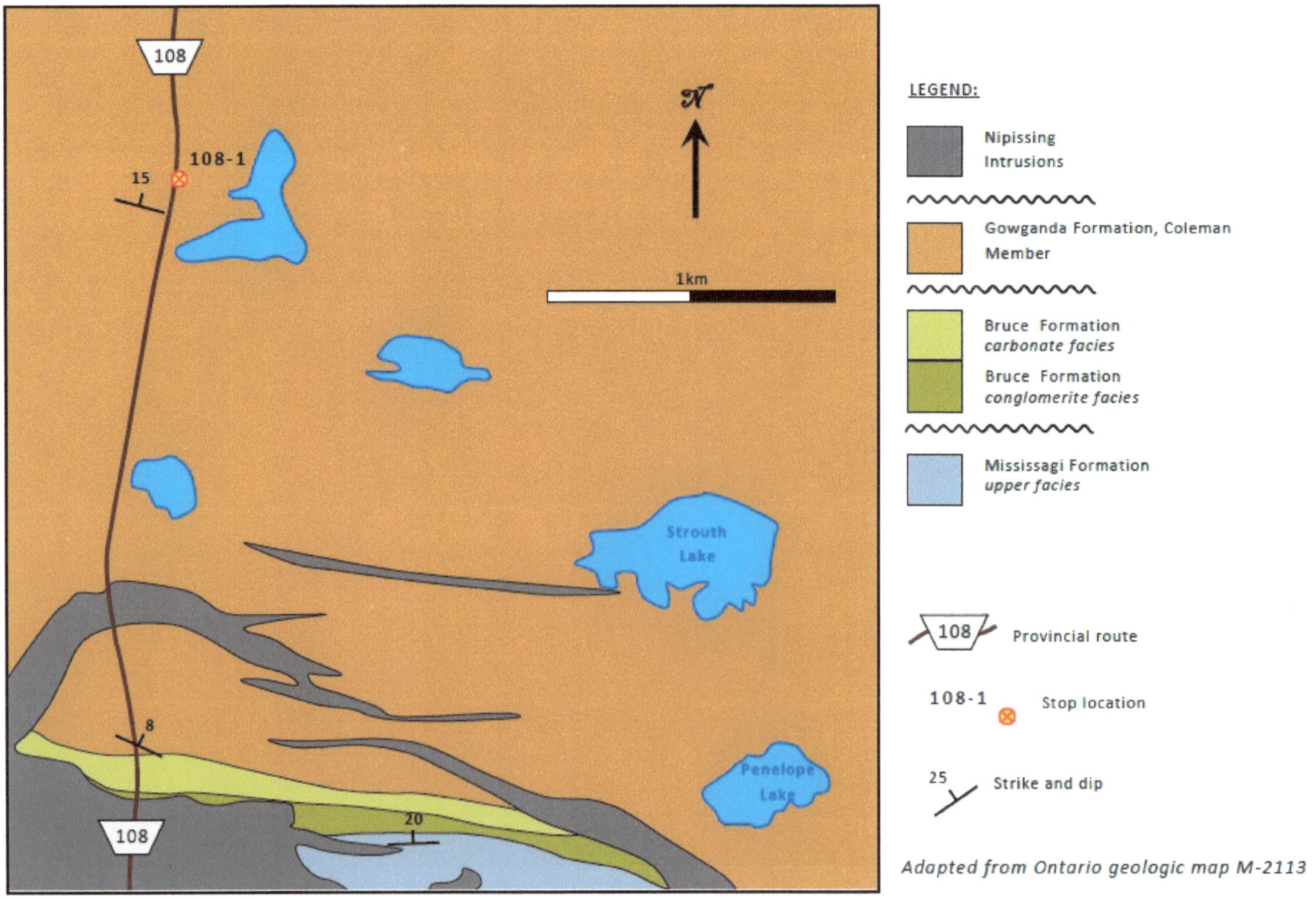

Adapted from Ontario geologic map M-2113

PHOTOS:

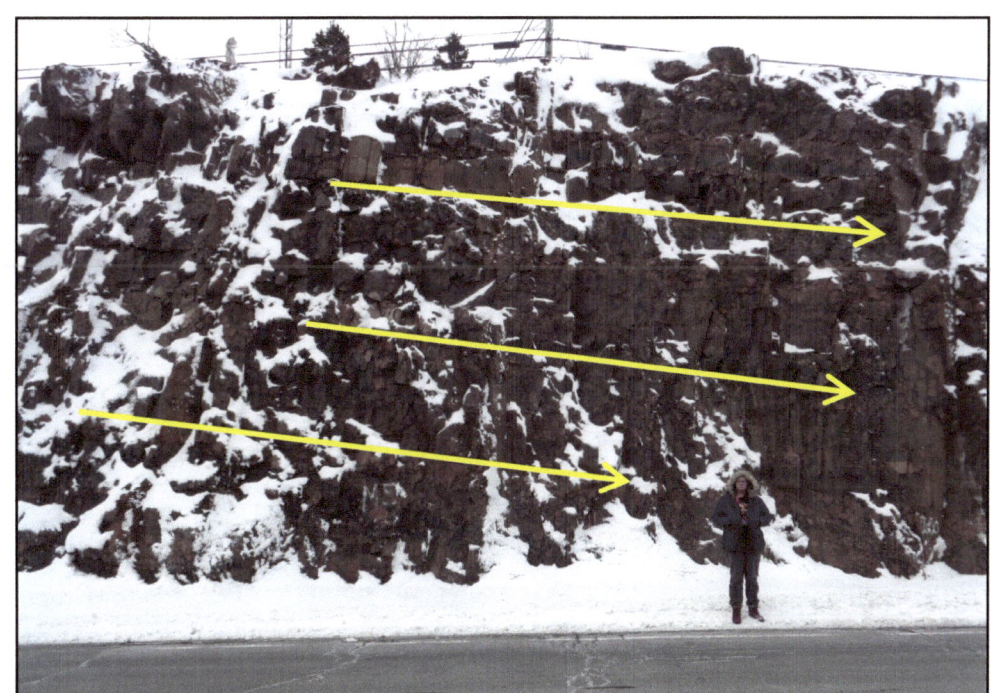

Outcrop, looking west. 5'10" (1.78m) person for scale. The yellow arrows are pointing in the direction of apparent dip, which is about half of actual dip.

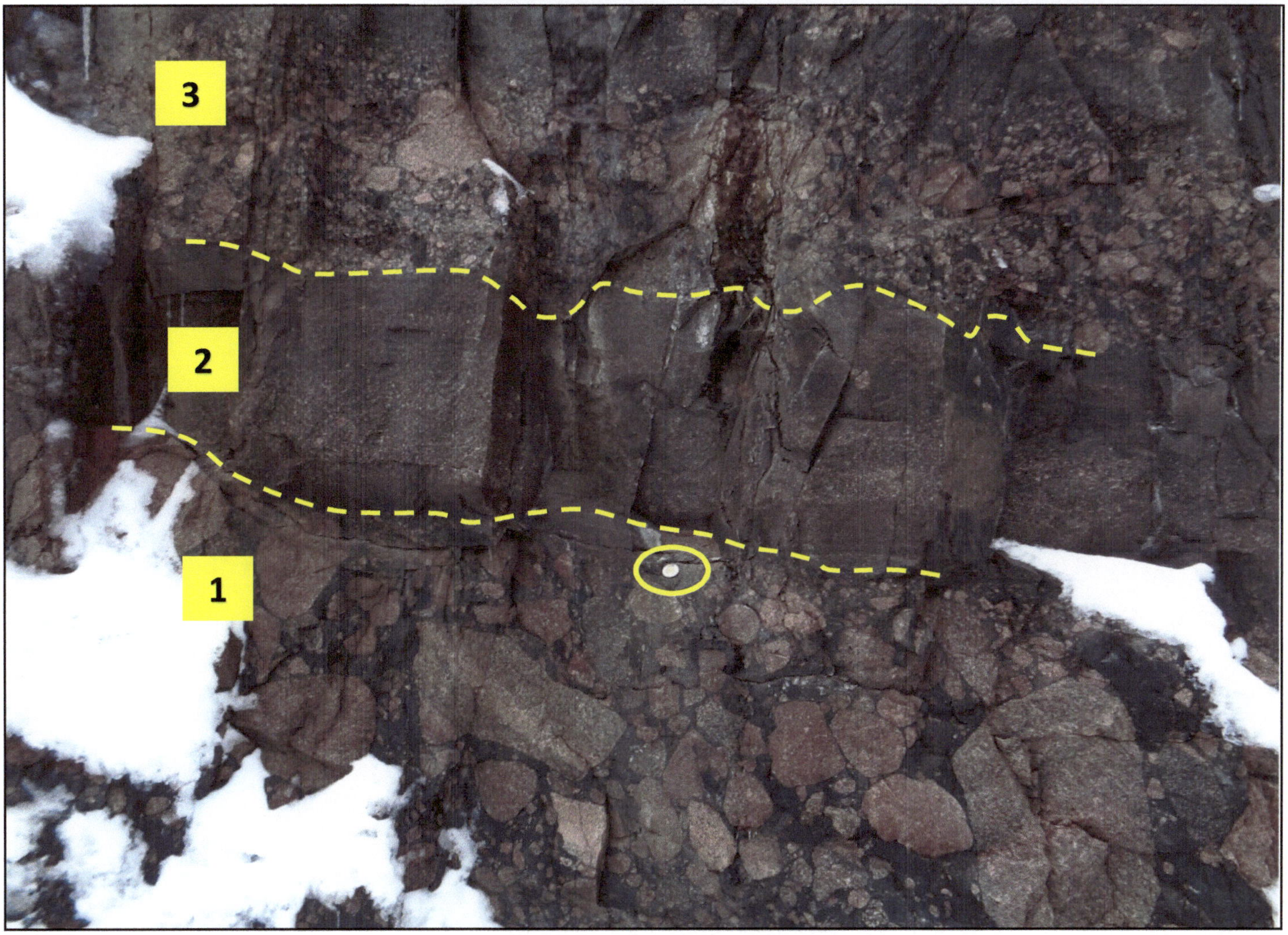

Outcrop, looking west. The different lithologies are numbered. 1) an unsorted, boulder supported, conglomerite (deposited directly by ice). 2) a pebbly, bedded, fining upwards proto-quartzite (low energy outwash). 3) Interbedded sorted sandy conglomerite and immature quartzite, that is internally gradational (outwash deposits with varying energy). $2 coin for scale (circled).

OUTCROP NAME: 640 meters South of Truck Terminal Road on ON-108 Outcrop **OUTCROP DESIGNATION:** 108-2

OUTCROP LOCATION: GPS: 46.41268 –82.66989 **ELEVATION:** 338.6m above MSL

FORMAL GEOLOGIC NAME: Bruce Formation (upper carbonate facies)

MAIN ROCK TYPE(S): Marble and dolomitic siltite

DESCRIPTION: Here we see the upper facies of the Bruce Formation. As where the lower part is glacial in origin (stop 546-6, p. 34-35) the upper part was likely deposited in a tidal flat environment. The upper part is a impure marble with mostly dolostone and dolomitic shale as its protolith. But that is not why I am including this outcrop.

Near the center of this outcrop is some structural geology in action (see photos on the next page). Here the consistent 21° to 25° north dip of the rock is interrupted by several small reverse faults. The faults all dip north and most change their dip as they propagate through the rock. Although geologists generally model faults as planar surfaces, this is more of what they actually look like in cross section. As you can see from the photos (next page), faults can merge into one fault further down. Faults are brittle failures in the rock (p. 6 and 16). What that means is the rock breaks instead of bends. There is also ductile deformation here. Ductile deformation occurs when the rock bends instead of breaks, as you can tell by some of the beds near the largest fault. The bedding dip changes direction near downthrown side of the biggest fault. We call these drag folds and they form during faulting. Even in brittle deformation, the rock is going to bend before it breaks and forms faults.

I included the strike and dips on the geologic map below, but not the faults; as they are too small to be mapped at the map scale.

FIGURE: Bedrock geologic Map

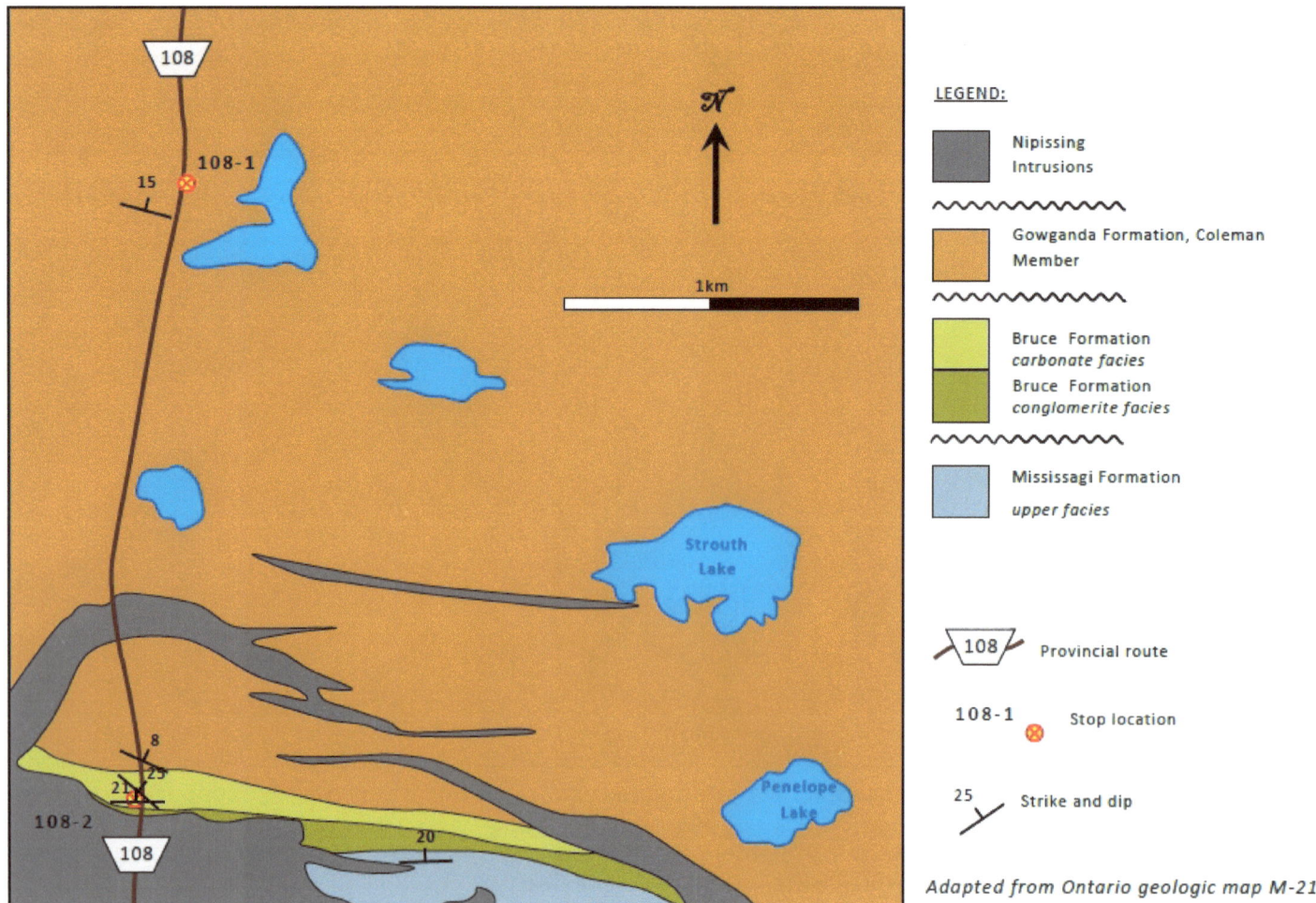

Adapted from Ontario geologic map M-2113

PHOTOS:

Outcrop, looking west.

Outcrop, looking west. I have highlighted the small reverse faults (solid lines), with relative movement (arrows). The dashed lines mark bedding. 25cm wide orange pouch for scale. See p. 23 for the raw photo.

OUTCROP NAME: Foodland Back Lot Outcrop **OUTCROP DESIGNATION:** 108-3

OUTCROP LOCATION: GPS: 46.38188 −82.64619 ELEVATION: 350.5m above MSL

FORMAL GEOLOGIC NAME: Mississagi Formation (lower facies)

MAIN ROCK TYPE(S): Proto-quartzite

DESCRIPTION: This outcrop is exposed in the back lot of a grocery called "Foodland" as of November 30, 2019 (dated of the photos on the next page). Within the city limits of Elliot Lake (see geologic map below), the Mississagi is the best exposed bedrock formation. Unlike stop 639-9 (p. 54-55), where we visited the upper facies, here we visit the lower facies.

Conglomerite is present locally, but it is not exposed at this outcrop. It does contain isolated pebbles. Here the rock is almost all medium crystalline proto-quartzite. It is olive gray in color, with red colors due too secondary rusting (post lithification). Although the rock looks massive on fresh surfaces, closer examination shows it is thickly cross laminated to thinly cross bedded. The cross bedding makes an exact strike and dip at this location hard to obtain. Based on surrounding measurements, the rocks have a modest dip of 15° to 20° north to northeast.

This part of the Mississagi was likely deposited in a nearshore beach to fluvial stream environment.

FIGURE: Bedrock geologic Map of Elliot Lake Ontario

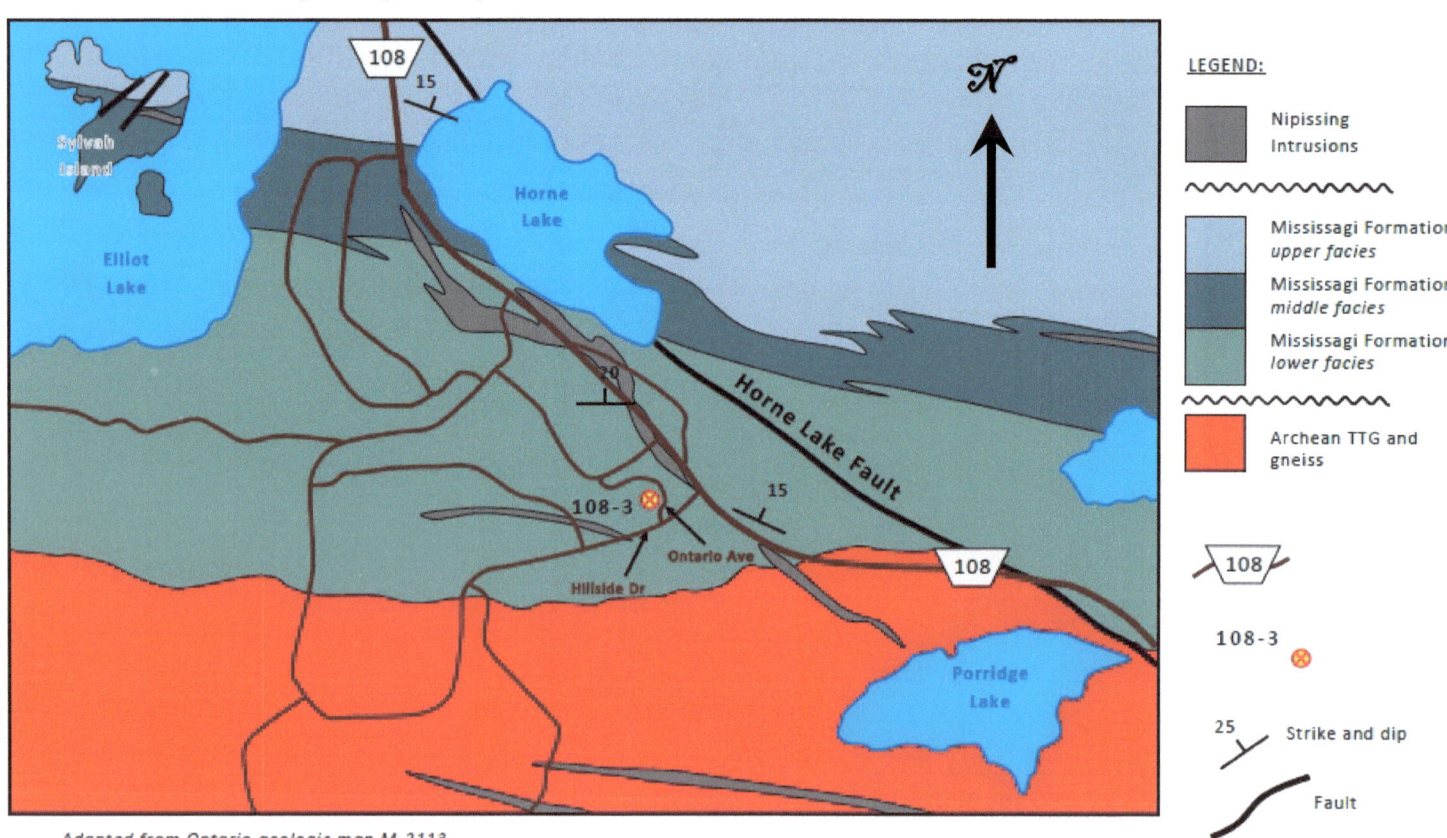

Adapted from Ontario geologic map M-2113

NOTE: The roads above are not all the roads in the town of Elliot Lake. They are just included so you can roughly orient yourself.

PHOTOS:

Outcrop, looking southwest. Automobiles for scale.

Outcrop, looking southwest. Black dashed lines highlight some cross beds. Arrows are pointing to cores drilled to remove rock. $2 coin for scale (upper center right).

OUTCROP NAME: 690 meters South of Nordic Lake Outcrop   **OUTCROP DESIGNATION:** 108-4

OUTCROP LOCATION: GPS: 46.37340 −82.59873   **ELEVATION:** 344.7m above MSL

FORMAL GEOLOGIC NAME: Archean TTG cross cut by Nipissing

MAIN ROCK TYPE(S): Tonalite-diorite cross cut by diabase

DESCRIPTION: This outcrop is on both sides of 108 but it is much taller on the north side. The host rock is light gray from a distance, but upon closer inspection it is mottled gray and light pink. It is relatively undeformed Archean tonalite-diorite. Rust often forms on the surface.

It is often cross cut by red felsic dikes (see photos on the next page). This outcrop holds some other things on the north exposure. It is cross cut by a big black Nipissing dike. The road cut isn't parallel to dip, so the true thickness is a lot thinner. You have an oblique view to true strike and dip. The true orientation of the dike is ~N75E83SE.

Near the east end of the outcrop there is a small normal fault with ~0.7m of displacement. The downthrown block is to the right (east).

There are two nearby closed mines to the northeast. The larger one is the Nordic Mine, where uranium was mined from the Mississagi Formation from 1957-1968 with 12 million tons of ore removed (Keen, 2006). The much smaller mine of Buckles was adjacent and south of the Nordic Mine. It only operated from 1957-1958. It was also mined for uranium (ontarioexplorations101.com/elliot-lake-ontario-mine/buckles-uranium-mine). Both mines were operated by "Rio Algoma Ltd."

FIGURE: Bedrock geologic Map

PHOTOS:

Outcrop, looking south-southeast., at the undeformed Archean TTG. $2 coin for scale

Red felsic dikes (yellow arrows), in the Archean TTG, looking north-northwest. No scale.

Black Nipissing dike intruding into the gray Archean TTG, looking northwest. This is an oblique view of the dike. It trends N75E83SE. Outcrop is about 5.5 to 7 meters high.

A red felsic dike (yellow arrows) split by a normal fault (black line and black half arrows show relative movement). Outcrop is about 6 to 7.5 meters high.

OUTCROP NAME: 43 meters on Algom Nordic Mine Road from ON-108 Outcrop

OUTCROP DESIGNATION: 108-5

OUTCROP LOCATION: GPS: 46.37095 –82.57969

ELEVATION: 351.1m above MSL

FORMAL GEOLOGIC NAME: Keewatin

MAIN ROCK TYPE(S): Foliated amphibolite

DESCRIPTION: This outcrop is on the west side of the road. The name "Algom Nordic Mine Road" is on the sign, but Google Earth lists it as "Nordic Trailer Park".

The rock is nearly black and highly metamorphosed. It is best described as a foliated amphibolite. Its protolith was likely volcanic.

The strike is roughly parallel to the contact with the Archean TTG and it dips away from the TTG. The trend is N64W65NE. This demonstrates how the TTG is younger than the Keewatin.

FIGURE: Bedrock geologic Map

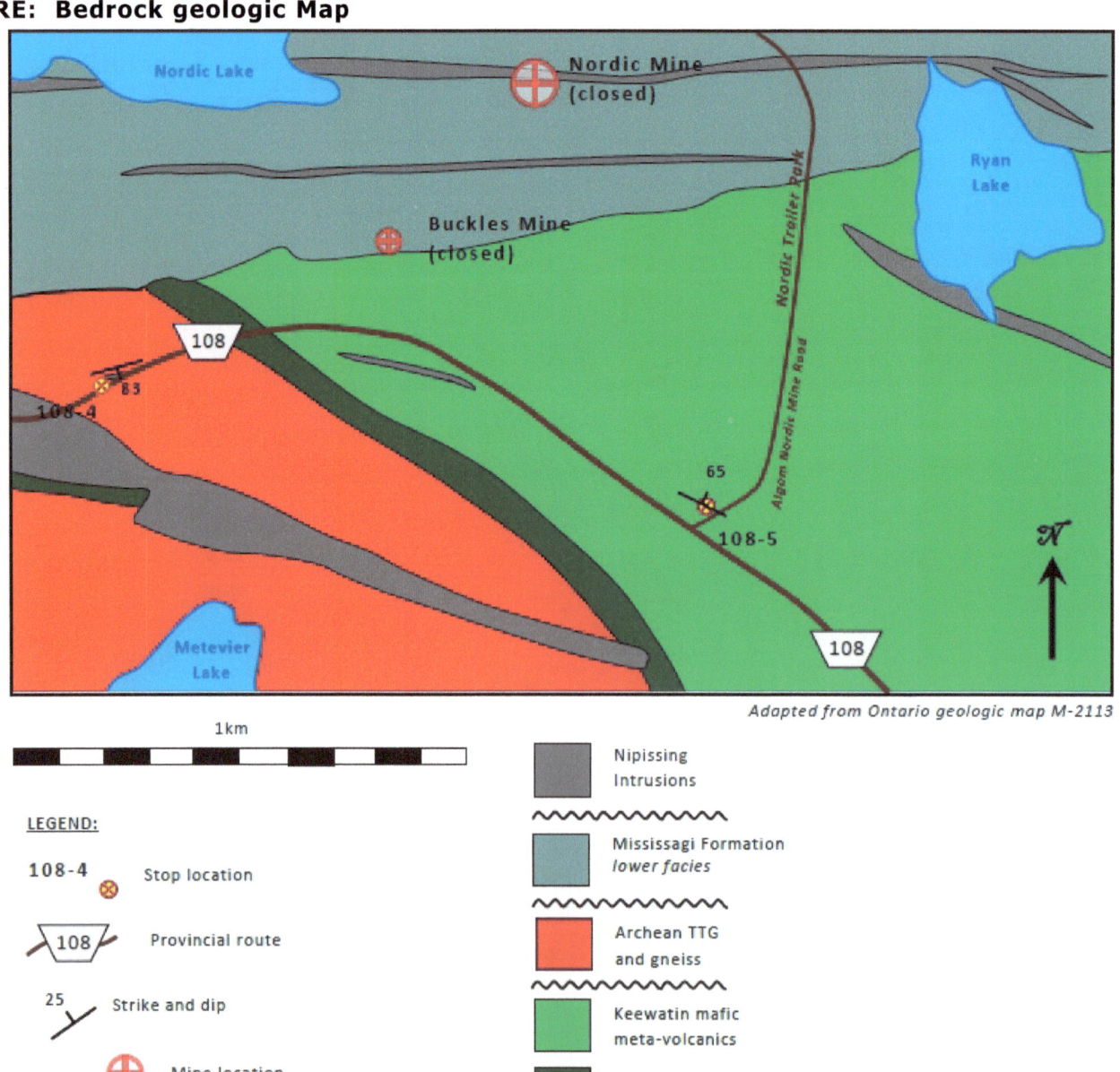

Adapted from Ontario geologic map M-2113

PHOTOS:

Outcrop, looking northwest. 5'10" (1.78m) person for scale.

Outcrop, looking northwest. $2 coin for scale (yellow circle). Notice the rusty surface caused by the weathering of the dark mafic minerals.

Outcrop, zoomed in on the amphibolite and looking northwest. $2 coin for scale.

OUTCROP NAME: 4710 meters North of Trans-Canada 17 on ON-108 Outcrop

OUTCROP DESIGNATION: 108-6

OUTCROP LOCATION: GPS: 46.25403 –82.56592

ELEVATION: 258.8m above MSL

FORMAL GEOLOGIC NAME: Archean TTG and Nipissing Intrusions with quartzolite

MAIN ROCK TYPE(S): Gneiss, diabase, and quartzolite

DESCRIPTION: This outcrop is on the east side of ON-108. The host rock is typical pink Archean gneiss that has a parent rock of tonalite-granodiorite. The foliation in the rock varies slightly but at this outcrop I measured it to be N25W57NE, but that isn't the coolest part of this outcrop.

The outcrop is cross cut by a black Nipissing diabase dike. Locally the Nipissing is diabase or gabbro. It Highly foliated near the contact with the gneiss.

The Nipissing has also been intruded by white quartzolite. Quartzolite is an igneous rock cooled from magma that is nearly 100% quartz. These are rare, but somewhat common around Lake Superior and the north shore of Lake Huron.

FIGURE: Bedrock geologic Map

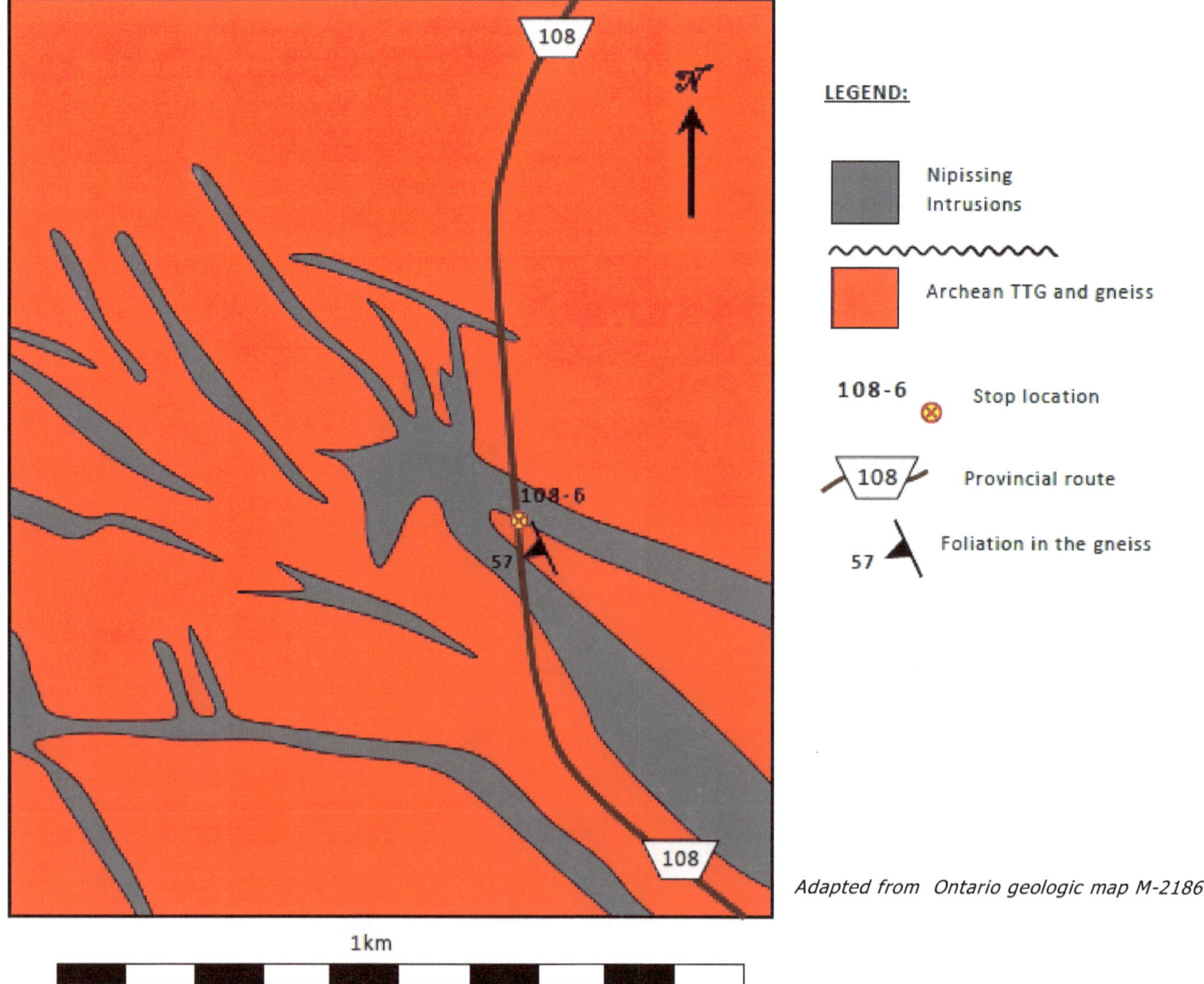

Adapted from Ontario geologic map M-2186

PHOTOS:

Outcrop, looking north. Truck for scale

Archean gneiss, looking northeast. $2 coin scale.

Outcrop, looking east. 15.2cm blue pen for scale at top center. The yellow lines divide the units. We know the Nipissing intrudes the gneiss because there is an altered zone (yellow arrow). The dotted yellow lines follow foliation in the diabase. There is also a quartzolite intruding the diabase, which maybe of magmatic origin. The red arrow points to another intrusion of quartzolite, which could be part of the larger one.

References

Baumann, S.D.J., 2019, *Stratigraphic Column and Ages of the Huronian Supergroup and the Chocolay Group*, Midwest Institute of Geosciences and Engineering (MIGE), Publication: G-082017-1C

Bennett, G., 2006, *The Huronian Supergroup between Sault Ste. Marie and Elliot Lake*, Institute on Lake Superior Geology (ILSG), Field Trip Guidebook, v.52, Part 4

Bouma, A.H., Ravenne, C., 2004, *The Bouma Sequence (1962) and the resurgence of geologic interest in the Frenceh Maritime Alps (1980s): the influence of Gres d'Annot in developing the ideas of turbidite systems*, Geological Society of London, Special Publications, v.221, p.27-38

Bowen, N.L., 1922, *The Reaction Principle in Petrogenesis*, Journal of Geology, v.XXX, no.3, p.177-198

Canfield, D.E., Ngombi-Pemba, L., Hammarlund, E.U., et al., 2013, *Oxygen dynamics in the aftermath of the Great Oxygenation of Earth's Atmosphere*, Proceedings of the National Academy of Sciences (PNAS), v.110, no.42, p.16736-16741

Cannon, W.F., Schulz, K.J., Horton Jr., J.W., King, D.A., 2010, *The Sudbury impact layer in the Paleoproterozoic iron ranges of northern Michigan, USA*, Geological society of America (GSA), v.122, no.1/2, p.50-75, doi: 10.1130/B26517.1

Card, K.D., 1977, *Stratigraphy, Sedimentology, and Petrology of the Huronian Supergroup in the Sudbury-Espanola Area*, Ontario Division of Mines, Geoscience Study 16

Card, K.D., 1978, *Geology of the Sudbury-Manitoulin Area, Districts of Sudbury and Manitoulin*, Ontario Ministry of Natural Resources, Ontario Geological Survey (ONGS), Report 166

Colivine, 2011, *Bowen's Reaction Series*, PNG diagram, public domain, Wikipedia

Davidson, A., 1995, *A Review of the Grenville orogen in its North American type area*, AGSO Journal of Australian Geology & Geophysics, v.16, no.1/2, p.3-24

Geissman, J.W., Bowing, S.A., Babcock, L.E. (compilers), 2018, *GSA Geologic Time Scale, V.5.0*, The Geological Society of America (GSA)

Gregory, J.W., and Barrett, B.H., 1927, *The stratigraphical position of the Keewatin*, Journal of Geology, v. 35, no. 2, p. 141*149

Heather, K.B., Desmond, M., Percival J.A., Blecker, M.W., 1195, *Tectonics and metallogeny of Archean crust in the Abitibi-Kapuskasing-Wawa region*, Geological Survey of Canada, Open File 3141

Hill, C., Corcoran, P.I., Aranha, R., Longstaffe, F., 2016, *Microbially induced sedimentary structures in the Paleoproterozoic upper Huronian Supergroup, Canada*, Earth Sciences, publication 9

Keen, J., 2006, *Uranium mining in Canada-past and present*, Mine watch Canada

Leith, C.K., Lund, R.J., Leith, A., 1935, *Pre-Cambrian rocks of the Lake Superior region*, United States Geological Survey, Professional Paper 184

Long, D.G.F., 1987, *Sedimentary framework of uranium deposits in the southern Cobalt embayment, Ontario, Canada,* in the IAEA-TECDOC-427 *Uranium deposits in the Proterozoic quartz pebble conglomerates*, p. 155-158

Mustard, P.S., 1985, *Sedimentology of the Lower Gowganda Formation, Coleman Member (Early Proterozoic) at Cobalt, Ontario*, National Library of Canada, Canadian Theses Service, Department of Geology, Carleton University in Ottawa, Ontario

References

Robertson, J.A., 1967, *Geology of the Spragge Area*, Ontario Department of Mines, Geology Branch, Open File Report 5010, 1966 Project 61-6

Rousell, D.H., Brown, G.H. (Editors), 2009, *A Field Guide to the Geology of Sudbury Ontario*, Ontario Geological Survey (ONGS), Open File Report 6243

Streckeisen, A. L., 1974, *Classification and Nomenclature of Plutonic Rocks. Recommendations of the IUGS Subcommission on the Systematics of Igneous Rocks*, Geologische Rundschau. Internationale Zeitschrift für Geologie. Stuttgart. Vol.63, p. 773-785.

Thompson, A.B., England, P.C., 1984, *Pressure-Temperature-Time Paths of Regional Metamorphism II. Their Inference and Interpretation using Mineral Assemblages in Metamorphic Rocks*, Journal of Petrology, v.25, Part 4, p.929-955

Winter, J.D., 2001, *An Introduction to Igneous and Metamorphic Petrology*, Text Book, published by Prentice Hall, ISBN: 0-13-240342-01

References (Maps)

Geologic Maps produced by the Ontario Geological Survey and Ministry of Mines:

M-2012: 1962, Abraham, E.M., Robertson, J.A.,
    Iron Bridge Area, District of Algoma, Ontario, Scale: 1:31680

M-2113: 1967, Abraham, E.M., Robertson, J.A.,
    Township 149, Algoma District, Scale: 1:15840

M-2114: 1967, Abraham, E.M., Robertson, J.A.,
    Township 150, Algoma District, Scale: 1:15840

M-2186: 1970, Abraham, E.M., Robertson, J.A.,
    Long and Spragge Townships and Part of Indian Reserve No.7, Scale: 1:31680

M-2306: 1975, Wood, J.
    Hembruff and Hughson Townships, Algoma District, Scale: 1:31680

M-2346: 1976, Robertson, J.A.,
    Poulin and Sagard Townships, Algoma District, Scale: 1:31680

M-2347: 1977, Robertson, J.A.,
    Nicholas and Raimbault Townships, Algoma District, Scale: 1:31680

M-2399: 1977, Siemiatkowska, K.M.
    Endikai Lake, Algoma District, Scale: 1:31680

M-2670: 2003, Johns, G.W., McIlraiths, S., Muir, T.L.,
    Precambrian Geology Compilation Series, Sault Ste. Marie, Blind River Sheet, Scale: 1:250000

www.ingramcontent.com/pod-product-compliance
Lightning Source LLC
Chambersburg PA
CBHW051201220526
45473CB00003B/864